PRODUCT STEWARDSHIP

Life Cycle Analysis and the Environment

PRODUCT STEWARDSHIP

Life Cycle Analysis and the Environment

Kathleen Sellers

CRC Press
Taylor & Francis Group
Boca Raton London New York

CRC Press is an imprint of the
Taylor & Francis Group, an **informa** business

CRC Press
Taylor & Francis Group
6000 Broken Sound Parkway NW, Suite 300
Boca Raton, FL 33487-2742

First issued in hardback 2019

ISBN-13: 978-1-4822-2329-3 (hbk)

Visit the Taylor & Francis Web site at
http://www.taylorandfrancis.com

and the CRC Press Web site at
http://www.crcpress.com

Révérence

To Bob Thoma and the other teachers and choreographers who trained me to be a dancer, and in so doing made me a better engineer.

Contents

Acknowledgments

Fred Lont, the ARCADIS librarian, and my research assistant, Alexis Palmer of Northeastern University, helped me to anchor this book in scientific data and applicable regulatory information. Their resourcefulness is greatly appreciated. Tim Richards and Andy Lewandowski of ARCADIS drafted many of the figures in this book; their work helped to make complex concepts clearer.

Many thanks to the scientists and engineers who reviewed the draft manuscript and provided helpful commentary: Tina Armstrong, PhD; Tina Berceli-Boyle, PE; James A. Hill; David Ludwig, PhD; Maryann Sanders; and Mike Thomas, CIH. Any remaining errors are my own.

Kathleen Sellers

About the Author

Kathleen Sellers, PE, is a vice president and principal environmental engineer with ARCADIS U.S., where she supports clients' product stewardship programs. Ms. Sellers earned a BS in chemistry from Indiana University and an MS in environmental engineering from the University of Massachusetts. She is currently on the board of directors of the Product Stewardship Society. This is her fourth book for CRC Press. Previous works include *Nanotechnology and the Environment* (2009), which she edited and coauthored.

1

Introduction

This book explores the global product stewardship of synthetic chemicals with a particular focus on the regulatory and scientific analyses of the life cycle consequences of bringing a chemical to market. That ossified sentence barely hints at the complexity of choices we make every day, choices that can have effects—or not—halfway around the planet. We humans make sense of the world through the stories we tell. And so, with rigorous attention to fact, this book expands on the bone-dry premise of its opening sentence by using case studies to explore two deliberately naïve questions about product stewardship, life cycle analysis, and the environment.

Infants born in the far north of Quebec, in the remote Nunavik region, have polychlorinated biphenyls (PCBs) in their umbilical cord blood [1]. Sediments off the Great Barrier Reef contain elevated concentrations of tributyl tin, a compound used to prevent ship fouling [2]. Scientists have found perfluorooctane sulfonate, used in applications as diverse as waterproof clothing and firefighting foams, throughout the world's oceans to depths of 1,500 meters or more [3]. These findings bring us to the first of our two questions: How is it that man-made chemicals are found in our bodies and in even remote parts of our planet?

It is another question entirely whether, to be blunt, such findings of man-made chemicals matter. Some take the philosophic approach that no amount of anthropogenic chemicals in the environment is acceptable. Others seek to balance the known effects of exposure to such chemicals against the economic and other benefits of their use. Consider that manufacturers used PCBs to cool and lubricate electrical equipment; these essentially fireproof compounds provided a measure of safety [4]. But PCBs are not safe in every context. Inuit children from the Nunavik region whose umbilical cord blood contained PCBs showed neurological damage in the form of poorer visual recognition memory [1]. The now nearly universal regulatory ban on the intentional manufacture of PCBs reflects such concerns about the hazards of exposure and shows how the answer to the question, "Does it matter?" can illuminate the answer to the question, "How can it be?" This book explores the value judgments we make about chemicals in the environment, as refracted through the prism of chemical control regulations.

All of us—legislators, regulators, manufacturers, and consumers—share responsibility for the thoughtful use of chemicals. Current conditions reflect conscious choices, whether or not those choices reflect an accurate prediction of the consequences. Such choices often embody risk management decisions. We must implicitly decide whether the possible consequences of manufacturing and using a chemical, insofar as we can predict those consequences, outweigh the benefits of the chemical's use.

As much as possible, this book provides a dispassionate view of the myriad factors that influence both our decision making and the consequences of those decisions. The objective of this study is to understand how current conditions have come to be and to help us to make wise decisions in the future.

1.1 Setting the Stage

Neither the industrial production of chemicals nor the need for stewardship is new. Isotopic analysis of ice core samples has shown, for example, that as early as 680 BC winds may have carried lead emissions from mining and smelting activities in Europe northward to Greenland [5]. The potential for pollution to occur grew with the rise of the chemical industry in the 1700s. Figure 1.1 shows some of the milestones in the evolution of the modern chemical industry [6]. From its origins in the production of acids and alkalis in the 1700s through current developments in nanotechnology, the chemical industry has continually brought new products to market. Those markets and their supply chains now span the globe.

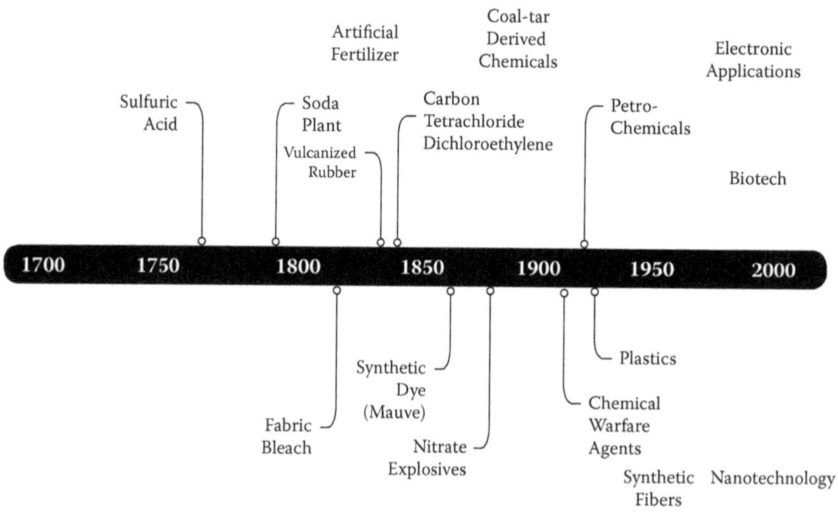

FIGURE 1.1
Selected milestones and trends in the modern chemical industry.

1.2 Technical Building Blocks

Product stewardship means "responsibly managing the health, safety, and environmental aspects of raw materials, intermediate, and consumer products throughout their life cycle and across the value chain in order to prevent or minimize negative impacts and maximize value" [7]. Chapter 2 of this book discusses the technical tools that a product steward uses to achieve this goal. These tools include the techniques to characterize and predict the fate and transport of chemicals in the environment upon their manufacture and use. The tools also include the methods used to calculate the possible risks to human health and the environment that may result. Chapter 2 also describes the formal process of life cycle assessment, which uses these tools to evaluate the potential effects on the environment as a result of the production, use, and disposal or recycling of a product.

1.3 Regulating Chemical Manufacture and Use

Chapter 3 describes the evolution of chemical control laws from their origin in the 1970s to the present. Such a broad topic requires focus to be intelligible, and so this book concentrates on the regulatory regimes in three of the major chemical-producing economies: the United States, European Union, and China. These regulatory schemes have, as a matter of practicality, distinguished between the chemicals in existence at the time regulation began and the chemicals created afterward. Figure 1.1 hints at the variety of substances that existed when chemical control laws originated in the 1970s. Chapter 3 describes the regulation of both existing and new chemicals, using examples to illustrate the complexity of the factors that affect their regulation as well as voluntary stewardship efforts by industry.

1.4 Telling the Story

Whereas Chapters 2 and 3 illustrate essential points with short examples, Chapter 4 tells the stories of four chemical products in more detail. These longer-form case studies integrate the technical and regulatory foundations of product stewardship and allow us to explore past risk-management decisions. Case studies of three chlorinated hydrocarbons show the effect of molecular structure on the behavior of a chemical in the environment and on the possible consequences of exposure. These case studies also show how our attitudes toward the use and regulation of chemicals have evolved over time, as has our view of the balance between risks and benefits. A fourth case study examines the effect of particle size on the behavior of a chemical-based product and how current regulatory regimes may apply.

1.5 Synthesis

The book concludes with a synthesis of the technical and regulatory factors that bear on the central questions of the book. Chapter 5 explores the themes of risk management and our ability to make thoughtful risk management decisions based on the state of the science.

Now let's begin.

References

[1] Boucher, O., Muckle, G., Jacobson, J.L., Carter, R.C., Kaplan-Estrin, M., Ayotte, P., Dewailly, É., and Jacobson, S.W. 2014. Domain-specific effects of prenatal exposure to PCBs, mercury, and lead on infant cognition: Results from the Environmental Contaminants and Child Development Study in Nunavik. *Environ. Health Perspect.* 122: 310–316.

[2] Negri, A. and Marshall, P. 2009. TBT contamination of remote marine environments: Ship groundings and ice-breakers as sources of organotins in the Great Barrier Reef and Antarctica. *J. Environ. Manage.* 90: S31–S40.

[3] Yamashita, N., Taniyasu, S., Petrick, G., Wei, S., Gamo, T., Lam, P.K.S., and Kannan, K. 2008. Perfluorinated acids as novel chemical tracers of global circulation of ocean waters. *Chemosphere.* 70: 1247–1255.

[4] Agency for Toxic Substances and Disease Registry. 2001. *Toxic Substances Portal - Polychlorinated Biphenyls (PCBs).* Available at: http://www.atsdr.cdc.gov/toxfaqs/tf.asp?id=140&tid=26#bookmark02 (accessed March 15, 2014).

[5] Rosman, K.J., Chisholm, W., Hong, S., Candelone, J.P., and Boutron, C.F. 1997. Lead from Carthaginian and Roman Spanish mines isotopically identified in Greenland ice dated from 600 BC to 300 AD. *Environ. Sci. Technol.* 31(12): 3413–3416.

[6] Aftalion, F. 2005. *History of the International Chemical Industry: From the "Early Days" to 2000*, second edition. Philadelphia: Chemical Heritage Foundation.

[7] Product Stewardship Society. 2014. *Core Competencies for the Product Stewardship Professional.* Available at: http://www.productstewards.org (accessed October 19, 2014).

2

Life Cycle Considerations

This chapter describes the technical tools that scientists and engineers use to evaluate the distribution of chemicals in the environment and to estimate the resulting risks. Those tools range across multiple disciplines, beginning with the chemical engineering concept of mass balance and encompassing the reactions—biological and chemical—and the climatological and hydraulic forces that influence the fate and transport of chemical substances in the environment. Understanding the consequences of exposure to those chemicals in the environment requires that this chapter also discuss the basic principles of ecological and human health risk assessment.

These tools can be assembled to assess the consequences of a chemical's use throughout its life cycle. Life cycle assessment is a formal discipline with its own frameworks and terminology. At its heart, it combines a mass and energy balance with an evaluation of environmental impacts associated with a product, process, or service [1,2]. In keeping with the central questions of this book, the discussions of life cycle assessment and component tools in this chapter focus on chemical substances and neglect energy usage and the consequences of that usage.

2.1 Basis for Life Cycle Assessment: Conservation of Matter

The last thing that many consumers likely consider when selecting a product from store shelves may be that the chemical substances in that product will ultimately enter the environment, perhaps by volatilizing to air or washing down the drain; even after any solid waste is tossed in the trash for landfilling or incineration, chemicals may yet enter air, water, or soil. But matter is neither created nor destroyed. Short of nuclear reaction, the chemicals we use follow this principle of conservation of mass. Life cycle analyses consider the fate and transport of chemicals from the source of raw materials through use of the product to ultimate disposal. Such assessments fundamentally begin with a mass or material balance.

2.1.1 Mass Balance

A mass or material balance is nothing more than the process of accounting for the conservation of mass, considering changes of phase, dilution, and chemical reactions that occur within a system. It is the fundamental tool for evaluating what happens to a chemical released into the environment.

An engineer performing a mass balance begins by defining the system and the basis for the calculation. The basis might be the mass or concentration of a key reactant, or its input (mass/time). She then models the fate of a chemical within that system using the following general equation [3]:

$$\text{Accumulation} = \text{Input} - \text{Output} + \text{Generation} - \text{Consumption} \quad (2.1)$$

The protocol certainly sounds simple enough. But it quickly becomes quite complicated, even for a single chemical. Within even a simple system, multiple reactions can occur: chemical substances can move between media, and the flow of water or air move matter across system boundaries. (In many but not all cases, one can ignore the generation of contaminants within environmental systems and focus on input, output, and consumption.)

Mathematical equations that model these processes enable us to calculate the concentrations of a chemical released to the environment in air, water, soil, and sediment. Application of these equations requires numerous simplifying assumptions about the nature of the process (whether equilibrium/steady state or not, ideal or not), the physicochemical properties of the substance and its reactivity, and the characteristics of the environment itself.

The rate and extent of these processes can change over time. A mass balance usually reflects one of two assumptions or their converse: the system is at equilibrium or the system is at steady state. Equilibrium is a state of dynamic balance—such as that which occurs when forward and reverse reactions are equal—where there is no impetus for change within the system. Steady state simply means that the condition being evaluated (which might be concentration in a certain phase or the flow rate, for example) is essentially unchanging over a specified time period.

We come back to this concept of mass balance after examining some of the most important fate and transport processes that affect a mass balance.

2.1.2 Environmental Fate and Transport

Multiple processes can affect the fate and transport of a chemical substance, each of which can depend not only on the physicochemical properties of the substance but also on the environment around it. In general terms, the processes include changes in state, biodegradation and bioaccumulation, and chemical reactions; advective transport can move a substance with wind or water within a localized area or even globally. We look at these processes individually before exploring through examining models and specific examples how the processes combine to determine a chemical's fate and transport in the environment.

2.1.2.1 Moving between Phases

Chemical substances can move between different phases in a system that is not at equilibrium. Modelers sometimes refer to the movement of chemicals between environmental "compartments" via such mechanisms, that is, between soil, water, sediment, and air.

A substance released to the environment can volatilize from water to air, or sublime from solid to vapor phase. It can also be washed out of the air with rainfall that deposits the substance on land or in surface water. Scientists characterize the tendency of a chemical substance to partition between air and water by its vapor pressure and solubility in water, or, in dilute solutions at equilibrium, by the Henry's law coefficient (which can be measured or calculated from the ratio of the vapor pressure to solubility at a specified temperature). The Henry's law coefficient is sometimes referred to as an air–water partition coefficient.

Chemicals in water can sorb to sediment or soil in a reversible process that reflects the attraction and adhesion of molecules to solids. (Less commonly considered in environmental mass balances, some air pollutants can sorb to particulates in the atmosphere.) The *n*-octanol/water partition coefficient (K_{ow}) of a substance provides a crude indication of the tendency to partition to solids from water; a high K_{ow} value indicates that a substance is hydrophobic/lipophilic and would tend to sorb to solids. More sophisticated tests determine a distribution coefficient (K_d) or adsorption isotherm to relate the concentration in solution to the concentration sorbed to solids. The sorption coefficient K_d is the ratio between the concentration of a chemical in soil to the concentration in water which is in contact with the soil. Normalized to the organic carbon content of the soil, this coefficient becomes K_{oc} ($K_{oc} = K_d/$fraction organic carbon in soil) [4].

Calculations that represent interphase transport must account for the difference between ideal and nonideal behavior. The substances in an ideal system precisely obey simplified laws describing their behavior. An ideal gas, for example, would obey the familiar ideal gas law,

$$PV = nRT \tag{2.2}$$

which relates pressure (P) and volume (V) to the amount of the gas (n, moles), Avogadros constant (R), and absolute temperature (T). But most gases deviate from this law due to the effects of intermolecular forces and the volume of the gas molecules themselves on the behavior of the gas [3]. Engineers account for such deviations from ideal behavior by including a fugacity term in their calculations.* For example, fugacity replaces pressure in the ideal gas law to characterize the behavior of a nonideal gas. Fugacity can be related

* For a more elegant and rigorous explanation of fugacity, see *Multimedia Environmental Models: The Fugacity Approach* [5].

to pressure through an empirically determined or estimated fugacity coefficient that accounts for deviations from ideal behavior. Fugacity is related to the chemical potential of a nonideal substance and is proportional to concentration. It characterizes the "escaping tendency" of the substance, or the impetus to move between phases. At equilibrium the fugacity in all phases is equal [5].

2.1.2.2 Reactions and Biological Interactions

Once a chemical is released to the environment, biotic and abiotic reactions may degrade or transform the substance.

Biodegradability fundamentally determines whether a chemical will persist in the environment. For most substances, biodegradation or hydrolysis followed by biodegradation is the most important process for degradation in the aquatic environment [6]. Many organic chemicals can biodegrade. Inorganic compounds do not generally biodegrade but may be transformed by biological reactions. For example, certain microorganisms can transform elemental mercury into methyl mercury, a highly toxic form of the substance [7].

Biodegradation may be aerobic or anaerobic. Aerobic degradation can occur when sufficient oxygen is available to serve as the terminal electron acceptor in microbial metabolism. In the absence of sufficient oxygen, certain microbes can utilize other electron acceptors such as nitrate, sulfate, carbon dioxide, or another organic substance.

Some compounds break down partially to form other organic compounds. Those degradation products may persist or undergo further degradation. Ultimate biodegradation, sometimes referred to as mineralization, means that a substance degrades to carbon dioxide, water, biomass, and inorganic substances such as ammonia.

Engineers typically model the progression of biodegradation over time as a first-order reaction, that is,

$$C_t = C_0 e^{-kt} \tag{2.3}$$

where C_t is the concentration of a substance at time t, C_0 is the initial concentration, and k is the rate constant. The half-life of a substance, which indicates the time required for a chemical to degrade to half its initial concentration ($C_t = 0.5\ C_0$), is often used to characterize the biodegradability or persistence of a substance.

An organic substance resists biodegradation in two general cases: if it is hydrophobic or if the chemical bonds in the substance cannot be readily broken. Each case is briefly described below.

The first case pertains to hydrophobic substances that are poorly soluble in water and tend to sorb strongly to soil or sediment. Such characteristics limit the availability of a compound for microbial degradation [8]. Consider,

TABLE 2.1
Common Measurements of Biodegradability

Type of Test	Characteristics
Ready Biodegradability	Aerobic degradation of concentrated solution measured by change in dissolved organic carbon, biochemical oxygen demand, or carbon dioxide. "Pass/fail" test: if substance meets criteria for success, it is expected to undergo rapid and ultimate biodegradation in the environment.
Simulation Tests	Test results indicate rate of aerobic degradation under environmentally relevant conditions in soil, aquatic sediments, surface water, or a sewage treatment plant. Degradation rates determined by monitoring radiolabeled carbon or chemical analyses.
Inherent Biodegradability	Assess whether chemical has any potential for biodegradation under aerobic conditions.
Anaerobic Biodegradation Screening Tests	Evaluate potential for anaerobic ultimate biodegradation of organic chemicals in heated digesters for anaerobic sludge treatment.

for example, the two chemically related compounds 1,4-dichlorobenzene (DCB; CAS 106-46-7) and hexachlorobenzene (HCB; CAS 118-74-1). 1,4-DCB has a water solubility of 80 milligrams per liter (mg/L). The logarithm of K_{ow} (log K_{ow}) is 3.44. HCB has a solubility of 0.006 mg/L and log K_{ow} of 5.73. These data show that HCB is more hydrophobic than 1,4-DCB. The half-life of each substance reflects this difference. The half-life of 1,4-DCB in surface water is approximately 37 days and that of HCB is 2.7 to 5.7 years. (Chapter 4 discusses these compounds further and provides references for the values cited here.)

In the second case, the bonds between the atoms in a molecule are simply too strong to break through unenhanced biologically mediated reactions. (In some cases, enhancing biodegradation within an engineered treatment system can effect degradation.) The substance 1,4-dioxane (CAS 123-91-1) provides an example of this case for poor biodegradability. 1,4-Dioxane is a cyclic ether characterized by a strong C–O bond [9]. Although it is miscible in water and has an estimated log K_{ow} of –0.27, it is not biodegradable [10].

Understanding the potential for a chemical substance to biodegrade is crucial to assessing life cycle risks from exposure and is embedded in many regulatory programs.* Consequently, scientists have devised various tests for biodegradability as shown in Table 2.1 [6]. Such testing can determine

* Although biodegradation is considered in most chemicals regulations, it actually forms the foundation for assessing a new chemical substance in Japan. A new substance must be tested for biodegradability; if it biodegrades, no further tests are required and the substance is not regulated. If it is poorly biodegraded, bioconcentration testing is required to determine whether toxicity testing is required. Furthermore, the toxicity of the products of biodegradation must also be assessed [11].

whether a substance can be biodegraded at all, and if so, the rate at which it degrades. Testing can also assess the extent of biodegradation.

In contrast to biodegradation, which can destroy an organic compound, bioaccumulation and bioconcentration allow a compound to accumulate within the tissue of living organisms. The two terms are not precisely synonymous. Bioconcentration refers to the uptake of a chemical substance by all routes other than ingestion, whereas bioaccumulation refers to the net uptake of a substance from the environment by respiration, ingestion, and dermal absorption from any source, including the ingestion of other organisms [12–14]. These two phenomena are characterized by the bioconcentration factor (BCF) and bioaccumulation factor (BAF), respectively, which represent the ratios between the concentration of the substance in the organism and in the environment. Biomagnification, a third phenomenon, refers to the process by which the tissue concentrations of a chemical in organisms at one trophic level exceed tissue concentrations in organisms at the next lower trophic level in a food chain as the result of bioaccumulation [15]. These processes are typically of greatest concern for hydrophobic/lipophilic compounds characterized by a relatively high value of K_{ow}.

In addition to these biologically mediated processes, substances can undergo abiotic chemical reactions in the environment. These reactions, which include hydrolysis, oxidation, reduction, photo-oxidation, and photolysis, can transform or degrade the original compound. These reactions are generally not as critical to the environmental transformation of organic substances in water, soil, or sediment as biodegradation. The most significant, hydrolysis, refers to the decomposition of a compound upon reaction with water. It is often characterized by a first-order rate constant or half-life, DT_{50} [16].

As we might expect, the mechanisms for the degradation of air pollutants differ from the predominant mechanisms for degrading contaminants in water, soil, or sediment. In the atmosphere, pollutants can react with hydroxyl radicals ($OH^•$) generated from the photocatalyzed reaction between oxygen and water. Oxidation of air pollutants by hydroxyl radicals is the most significant environmental transformation reaction for some persistent organic pollutants [60].

The concepts of persistence (or resistance to degradation) and bioaccumulation are so critical to thinking about the life cycle of chemicals that regulators and other authorities pay them particular attention. A detailed discussion of the regulations can be found in Chapter 3; however, it is useful to consider in this discussion of environmental fate and transport some of the benchmarks by which chemicals are judged to be persistent and bioaccumulative.

The Stockholm Convention on Persistent Organic Pollutants, an international treaty, defines a chemical as persistent if its half-life in water is greater than two months, or the half-life in soil or sediment is greater than six months; additionally, a chemical is considered to have the potential for long-range transport in air if its half-life (in air) is greater than two days. A high potential for bioaccumulation is indicated by a BCF or BAF in aquatic species greater than 5,000 (unitless) or alternatively by a log K_{ow} value above 5 [17].

The European Union regulation on Registration, Evaluation, Authorisation and Restriction of Chemicals, known as REACH, reflects the following criteria: An organic or organometallic substance is said to be persistent if the half-life is greater than 60 days in marine water, greater than 40 days in fresh or estuarine water, greater than 180 days in marine sediment, or greater than 120 days in fresh or estuarine sediment or soil. Similarly, a substance is said to be bioaccumulative if the BCF is greater than 2,000 liters per kilogram (L/kg). The European Chemical Agency, or ECHA, has also established a screening criterion for bioaccumulation. If the log K_{ow} of an organic compound is greater than 4.5, then the compound may tend to bioaccumulate by passive diffusion driven by hydrophobicity. Below that threshold value, one can infer that the substance has an insufficient affinity for the lipids of an organism to exceed a BCF value of 2,000 L/kg. This screening criterion has several limitations. It does not apply to organic substances that may bioaccumulate by other mechanisms, or to metals or surface active compounds [18].

The following examples illustrate the principles of persistence and bioaccumulation. During the writing of this chapter, an estimated 7,500 gallons of 4-methyl-cyclohexane-methanol (MCHM; CAS 34885-03-5) spilled through a hole in a storage tank into the Elk River in West Virginia in the United States. From there, MCHM entered the public water supply of approximately 300,000 people. The persistence and potential bioaccumulation of this substance as well as its toxicity would logically shape the consequences of this spill. News reports at the time of the spill pointed out that authorities knew little about the toxicity or behavior of this substance in the environment [19,20]. For example, no data were publicly available in the Toxic Substance Control Act (TSCA) Test Submissions database maintained by the US EPA [21]. (In Chapter 3 we explore why little information may be available in the United States regarding the toxicity or behavior of an existing chemical substance.) Some limited data were available from the scientific literature, however. In a five-day biodegradation screening test seeded with acclimated sewage, 94% of the mass degraded [22]. In addition, MCHM has a log K_{ow} of 1.82 [23]. These data suggest that despite the immediate and urgent concerns about the public health and environmental consequences of the spill, MCHM would not persist in the environment or bioaccumulate.

In contrast, consider the characteristics of HCB. The global Stockholm Convention identifies HCB as a persistent organic pollutant (POP). (We revisit the story of HCB in Chapter 4 after looking briefly here at characteristics that influence its environmental fate and transport.) The half-life of HCB is estimated to range from 2.7 to 5.7 years in surface water and from 5.3 to 11.4 years in groundwater [24]. Laboratory experiments have determined BCF values for fish of up to 21,900 [25]. These characteristics help to explain why, in the words of the United Nations Environmental Programme (UNEP), HCB can "remain intact in the environment for long periods, become widely distributed geographically, accumulate in the fatty tissue of humans and wildlife, and have harmful impacts on human health or on the environment" [26].

This discussion has focused on chemical properties but other variables can influence the fate and transport of a chemical in the environment. The processes of phase change, biodegradation, bioconcentration/bioaccumulation, and abiotic chemical reactions are affected by environmental variables that include temperature, residence time within a system, degree of mixing, pH, the presence of oxygen, sunlight, or other chemicals, and the presence of living organisms. The infinite variability of natural environments certainly complicates efforts at estimating a mass balance on a substance released to the environment.

2.1.2.3 Transport: Local and Global

Advection, which refers to movement with flowing groundwater, surface water, or air, can transport a chemical substance through the environment. That dry technical definition only hints at the dramatic global scale and consequences of such transport. Let's begin to understand the mechanisms and scale of advection by looking at the water cycle.

2.1.2.3.1 Water Cycle and Ocean Currents

Water cycles continually through our environment. The movement of water within a region or more broadly within the world's oceans can carry contamination far from its point of origin, as we see in an example to follow.

Water evaporating from the earth's surface falls back to the ground as precipitation. Rainfall and snowmelt can soak into the ground, to be intercepted by the roots of plants or remain in the earth as groundwater until the flowing groundwater discharges to surface water or is pumped to the surface from a well. Water on the ground surface can run off into surface water, perhaps into a stream that conveys water to a larger river and then to the ocean.

Once in the ocean, the tides and currents driven by wind control the circulation of shallow water. Seven major currents, shown in Figure 2.1 [27], move water around the globe: the West Wind Drift (or the Antarctic Circumpolar Current), East Wind Drift, the North and South Equatorial currents, the Peru Current, the Kuroshio Current, and the Gulf Stream. These currents can move quickly. The Gulf Stream, for example, usually travels at a speed of 3 or 4 knots, which is equivalent to 5.6 to 7.4 kilometers per hour [27]. As these currents spiral through the ocean they form five major gyres: the North Atlantic, South Atlantic, North Pacific, South Pacific, and Indian Ocean gyres.

Deeper yet, water may enter the "Great Ocean Conveyor Belt" (Figure 2.2 [28]) where thermohaline circulation driven by variations in temperature and salinity affects the three-dimensional movement of the ocean's waters around the earth [29, 30]. As Arctic air chills the oceans near Iceland and sea ice forms, the salinity and density of the water at the surface increase

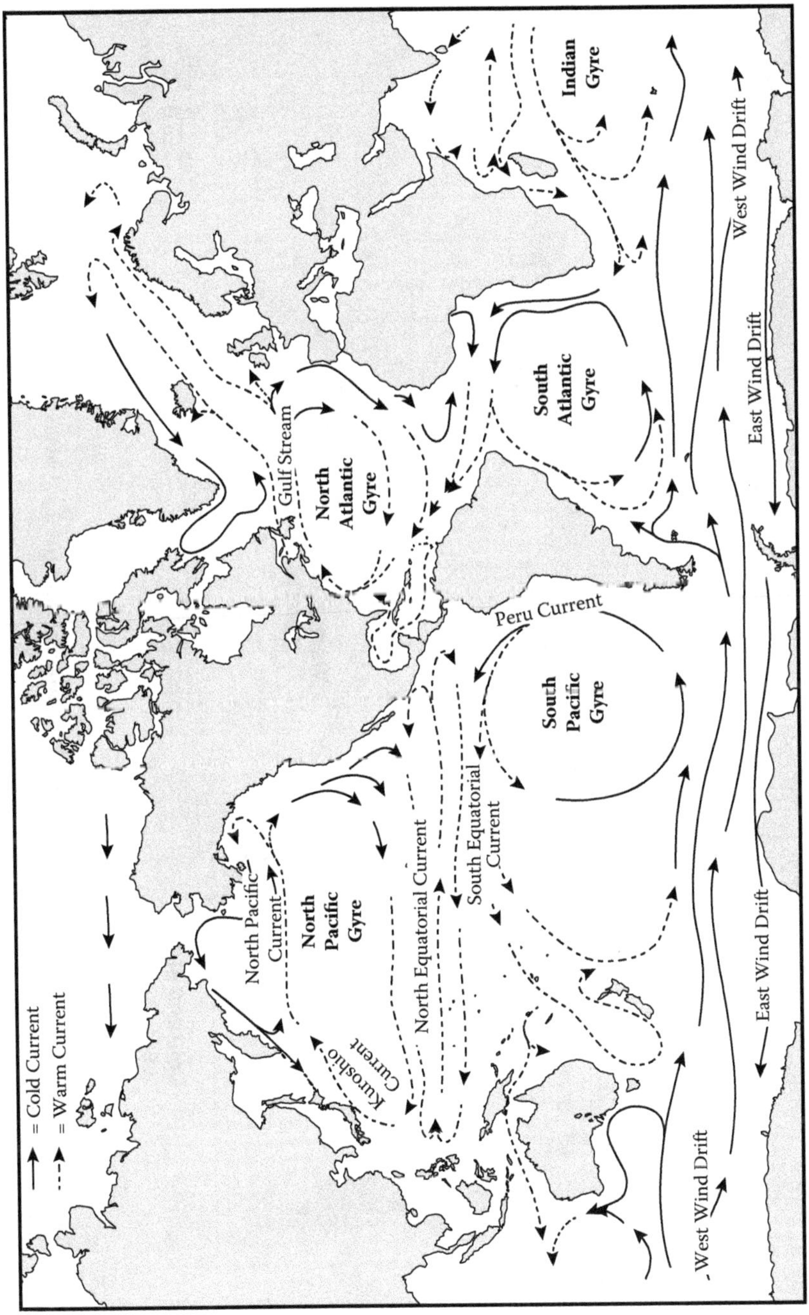

FIGURE 2.1
Ocean currents and gyres.

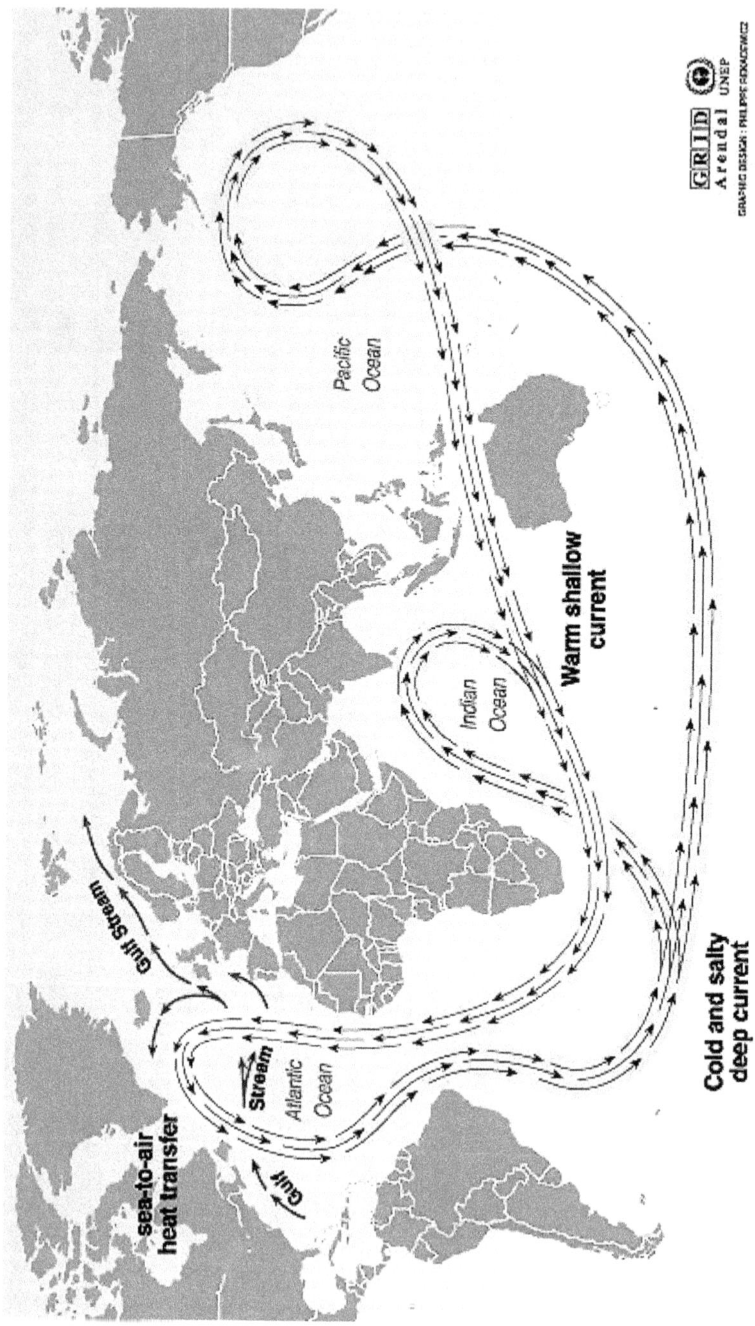

FIGURE 2.2
Great Ocean Conveyor Belt. (From Philippe Rekacewicz, UNEP/GRID-Arendal (http://www.grida.no/graphicslib/detail/world-ocean-thermohaline-circulation_57ea). With permission.)

until the cold water sinks. This mass, known as the North Atlantic Deep Water, flows slowly to the south deep in the Atlantic Ocean. Near the coast of Antarctica this mass mixes with water from the Antarctic and deep water circulating from the Pacific and Indian Oceans. Flowing to the east around Antarctica the mass cools, sinking further, and then splits in two. As shown in Figure 2.2, one stream moves north into the Indian Ocean and the other farther east into the Pacific Ocean. As these streams move north and east the water begins to warm, becoming less dense and upwelling toward the surface. Then the currents loop to the south and the west again into the South Atlantic. Water then flows north to the North Atlantic where it enters the conveyor belt again.

Water moves slowly through the Great Ocean Conveyor Belt. According to one estimate, it takes any given cubic meter of water approximately 1,000 years to complete the journey. Slow as it is, the sheer volume of water moving through this conveyor—estimated at more than 100 times the flow of the Amazon River [31]—dramatically affects the global ocean nutrient and carbon dioxide cycles. And as we explore in an example below, this global cycling has the potential to carry chemical contaminants far from their point of origin.

Human activities can release chemical substances at several points in the water cycle. Leachate from an unlined landfill or sewage treated in a septic tank may infiltrate groundwater. Treated wastewater commonly discharges to surface water. And air pollutants enter the water cycle when they dissolve in precipitation and return to earth.

Each step in the water cycle has the potential to convey chemical substances through the environment by advection. Precipitation can wash chemical substances out of the air and carry them to soil or surface water. On land, chemicals may then re-evaporate or infiltrate groundwater, or run off to surface water either sorbed to particulates or dissolved in water. Once in surface water, compounds may evaporate or be carried far downstream with flowing surface water, even to remote marine environments.

The mechanisms of degradation and sorption provide a counterpoint to these transport processes. Substances in solution may biodegrade or undergo chemical reactions that remove them from the environment. Hydrophobic compounds tend to sorb to soil or sediment (to a degree characterized by K_d or K_{oc}), and their transport through the environment is consequently retarded or attenuated. Such "reservoirs" of persistent compounds in surface soils can serve as long-term, low-level sources of contaminants in the environment [32]. Sediments may also serve as reservoirs of persistent compounds, or, when resuspended into the water column, contribute to their transport.

One research team questioned how to gauge which compounds had the highest potential for long-range transport in water [33]. They began with the criterion under the Stockholm Convention that a chemical may have the potential for long-range transport by air if its half-life in air is greater than two days, and scaled that factor to apply it to transport by water based on the relative velocities of long-range air and water flows. Their calculations suggest that in a river system a substance with a half-life between 8 to 15 days could travel a distance of 700 kilometers (km). In the ocean, considering the speed of Arctic currents, a half-life of 40 to 130 days would allow for a similar transport distance. Chemical compounds with half-lives greater than these calculated values could pose a potential for long-range transport in water. Although these screening criteria are not absolute, they do give a sense of the possibility that the flow of water may transport even degradable substances long distances.

Consider, for illustration, the fate and transport of certain perfluorinated compounds. Perfluorooctane sulfonate (PFOS; CAS 1763-23-1) is one of a large group of related perfluoroalkyl sulfonate substances generally described as having the molecular formula: $C_8F_{17}SO_2Y$, where Y = OH, a metal or other salt, halide, amide, and other derivatives including polymers. Most PFOS-related substances are high-molecular–weight polymers. In the environment or when metabolized by an organism, these polymers can degrade to PFOS. PFOS and related compounds have had many uses, including electric and electronic parts, fire-fighting foam, photo imaging, hydraulic fluids, and textile treatments (e.g., waterproofing or stain repellency), although some uses have been phased out [34].

The potassium salt of PFOS has a water solubility of 680 mg/L (at ~25°C) and a vapor pressure of 3.31×10^{-4} Pa. As a result of this low vapor pressure, PFOS is not expected to volatilize [34]. (Some PFOS-related compounds are more volatile, however.)

The strong carbon–fluorine bonds in PFOS resist degradation. Hydrolysis tests showed a half-life of approximately 41 years; the indirect photolytic half-life of PFOS has been estimated to be greater than 3.7 years; and, as reported by UNEP, repeated tests of biodegradability have failed to demonstrate any signs of biodegradation [34]. UNEP also reports that PFOS is highly bioaccumulative and can biomagnify, albeit not by the "typical" mechanism of partitioning into fatty tissues. Instead PFOS bioaccumulates by binding to plasma proteins and to proteins in the liver [34]. Due to this persistence and potential to bioaccumulate, PFOS is considered to be a POP under the Stockholm Convention.

PFOS may enter domestic sewage as a result of washing treated fabrics or cleaning carpets, for example. Industrial sewage may also contain perfluorinated compounds from the use of industrial additives, surfactants, and coatings. Conventional wastewater treatment would not degrade PFOS, although it might degrade related compounds to form PFOS. PFOS could sorb to the solids in the wastewater treatment plant or discharge to surface water in the effluent. Table 2.2 illustrates the global scope of such discharges. These data were generally obtained from plants that treated domestic sewage alone or in combination with industrial flows, unless otherwise noted.

Such wastewater effluents commonly discharge to surface waters, which ultimately discharge to the ocean. Various research teams have quantified

TABLE 2.2
Examples of Reported PFOS Discharges from Wastewater Treatment Plants

Location [Reference]	Number of Wastewater Treatment Plants: Detected/Studied	Maximum Concentration PFOS Detected in Effluent (ng/L)
United States (New York State) [35]	6/6	68
Canada [36]	6/7	72
Greece [37]	2/2	12.7
Spain [38]	16[a]	83.1[a]
Germany [39]	7/9	82.2
India [40]		12.0[b]
Republic of Korea [41]	17/18[c]	8.9
Japan [42]	6/8	207.5
Thailand [42]	5/5[d]	432.2
Vietnam [43]	0/1	< 0.8
China (Beijing) [44]	7/7	12.1
Australia [45]	2/2	3.7

[a] Mean concentration detected over two-year period.
[b] Untreated sewage.
[c] Included 15 municipal and 3 industrial wastewater treatment plants; some plants used advanced treatment processes.
[d] Industrial treatment plants.

PFOS in rivers and seas; the most sweeping studies have examined the distribution of PFOS in ocean waters.

One series of studies traces the occurrance of PFOS from upstream sources to the ocean. The data reported in Table 2.2 for the effluent from wastewater treatment plants in Germany were collected in 2007 from plants that discharged to the River Elbe in an industrialized area. Effluent from seven of nine plants contained detectable PFOS, at concentrations between 0.5 and 82.2 nanograms per liter (ng/L). The research team also collected and analyzed samples from 15 locations in the River Elbe. Those surface water samples reflected the contribution from municipal wastewater effluents and from industrial wastewater effluents or from surface runoff. All of the river water samples contained detectable PFOS, at levels between 0.5 and 2.9 ng/L [46]. In a second sampling campaign, this team collected samples from the German coast near the discharge of the River Elbe to the North Sea and from the open North Sea [47]. Three water samples from the mouth of the River Elbe contained 4.09–6.06 ng/L PFOS; 16 water samples from the North Sea along the German coast contained up to 2.26 ng/L PFOS; and three samples from the open North Sea contained up to 0.07 ng/L PFOS. (Not all of the samples collected off the German coast or in the North Sea contained detectable PFOS.) The research team concluded that the concentrations of perfluorinated compounds were highest close to industrial or highly populated areas, perhaps due to contributions from wastewater, but decreased rapidly in open seawater. Sorption to particulates, based on data not reported here, would account for only a fraction of the PFOS mass. A separate research team found similar results, detecting 8 ng/L PFOS in water samples collected from the mouth of the River Elbe and 1.8 ng/L PFOS in samples from coastal stations [48].

Researchers have ventured even farther asea to measure the levels of PFOS and related compounds in the waters of the world's oceans. They have detected PFOS throughout the ocean to depths of 1,000 m or more. Samples from coastal areas off Korea, China, Hong Kong, Japan, Germany, and England have contained PFOS. Farther out at sea, researchers have detected PFOS in samples from throughout the Pacific Ocean and Atlantic Ocean, and from the Southern (Antarctic) Ocean and Arctic Ocean. PFOS have also been detected in the Greenland Sea, Labrador Sea, Indian Ocean, and the South China Sea. The detected concentrations are generally on the order of tens of picograms per liter (pg/L; 1 pg/L = 0.001 ng/L), although concentrations of up to 291 pg/L have been detected in the open ocean [49–53].

Researchers have hypothesized that advective transport via surficial currents could account for the distribution of PFOS in shallow waters. Concentration gradients observed as samplers moved away from industrialized areas in Northern Europe toward the South Atlantic Ocean and followed known ocean currents, for example [50].

PFOS has also moved deeper in the oceans. In one study between 2002 and 2006, a team of researchers collected samples from 62 locations offshore of Japan and Angola and in the open ocean at various depths [53]. These open-water sampling locations were in the North Atlantic Ocean, Mid-Atlantic Ocean, central to western Pacific Ocean, south Pacific Ocean, South China Sea, Japan Sea, and Indian Ocean. The research team concluded that the distributions of PFOS and the related compound perfluorooctanoate (PFOA), laterally and vertically, were consistent with transport via the Great Ocean Conveyer Belt. The relatively low density distribution of samples and the relatively short period of use of PFOS (about 50 years) make it difficult to reach a definitive conclusion, but the results suggest, in the researchers' view, the potential for global three-dimensional transport of persistent compounds.

One can read the research reports described above with a technical fascination for the advances in analytical chemistry that enable scientists to monitor such remarkably low concentrations. Or a reader might be intrigued with the explanation that ocean currents can transport a persistent contaminant far from its point of origin over a period of decades or centuries. More to the point, these data support a simple hypothesis: because PFOS is highly soluble and persistent, ocean waters [54] and potential sorption to sediments can serve as the ultimate sink for PFOS with consequent exposure by aquatic organisms.

Let's return to the data in Table 2.2 to consider the possible worst-case consequences of finding PFOS in surface water. The levels reported in Table 2.2 represent the concentrations in wastewater effluent and might be further diluted upon discharge to surface water (as illustrated by the discussion of data from the River Elbe and other surface water bodies). Even without such dilution, and considering that the tabulated values were the maximum concentrations determined in each study, the data in Table 2.2 compare favorably with the results of aquatic toxicity testing. The UNEP risk profile [34] cites the following data: 42-day no observed effects concentration (NOEC) of 0.3 mg/L based on the mortality of fathead minnows (*Pimephales promelas*) [55]; a 10-day NOEC of 0.0491 mg/L for the growth and survival of the aquatic midge (*Chironomous tentans*) [56]; and a 96-hour NOEC value for algae (*Pseudokirchnerilla subcapitata)* of 5.3 mg/L [57]. These data indicate that toxic effects would not be seen in these aquatic organisms unless the concentration of PFOS was at least 100 times higher than the maximum values detected in wastewater effluents.

But does that mean that there is no cause for concern? Low concentrations discharged to the environment in wastewater might represent significant mass accumulation over time when one considers that the effluent from municipal sewage treatment plants, typically measured in units of million gallons per day, discharges all day, every day. Recall that PFOS does not biodegrade, can bioacccumulate, and has the potential to biomagnify. When the UNEP identified the PFOS family of compounds as POPs under the Stockholm Convention, scientists recognized that the convective transport of PFOS via water and airborne transport and then degradation of its more volatile cousins could have global consequences, saying [34]:

> Most notable and alarming are the high concentrations of PFOS that have been found in Arctic animals, far from anthropogenic sources. PFOS has been detected in higher trophic level biota and predators such as fish, piscivorous birds, mink, and Arctic biota. Also, predator species, such as eagles, have been shown to accumulate higher PFOS concentrations than birds from lower trophic levels. Even with reductions in manufacturing of PFOS by some manufacturers, wildlife, such as birds, can continue to be exposed to persistent and bioaccumulative substances such as PFOS simply by virtue of its persistence and long-term accumulation.

As this example illustrates, transport by water can be the dominant process for compounds with a long half-life in water and a tendency not to partition into air [33]. What of transport by air?

2.1.2.3.2 Atmospheric Transport

The water cycle is easy to visualize—at least for those of us not in arid climates—but the global mechanisms for atmospheric transport are perhaps less familiar. Understanding those mechanisms starts with the way in which scientists describe the atmosphere and the movement of air within it.

Meteorologists represent the earth's atmosphere as a series of five concentric layers. Closest to the earth is the troposphere. This layer, which varies between 8 and 18 km thick at different points on the earth, comprises the planetary boundary layer (i.e., the 1 to 2 km closest to the earth) and the

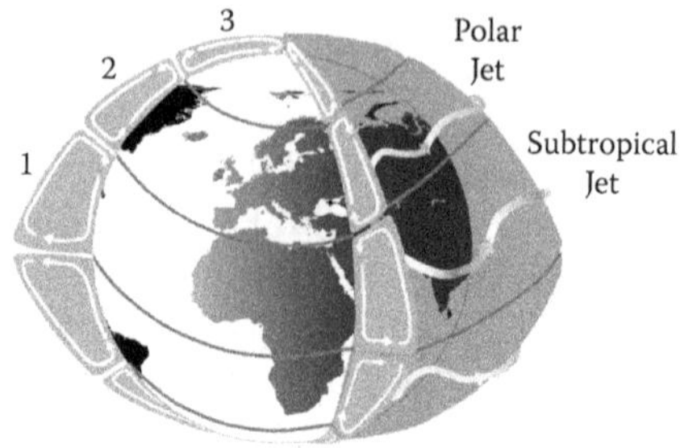

FIGURE 2.3
Air circulation in the troposphere: convection cells. 1) Hadley Cell, 2) Ferrel Cell, 3) Polar Cell.

free troposphere. The tropopause is the outer boundary of the troposphere. It separates the troposphere from the next layer, known as the stratosphere.

Air circulates within the troposphere from the equator toward the two poles. In simple terms, the hot air at the equator rises and moves toward the poles; when it encounters cold temperatures at the poles, it becomes more dense and sinks, then circulates back toward the equator. In fact, this occurs in a series of convection cells that run from the equator to each pole as shown in Figure 2.3 (adapted from [58]). The Hadley cells are located at the equator. Air circulates in the Hadley cells toward each of the poles, to a latitude of approximately 30° N/S. North and south of the Hadley cells lie the Ferrel cells. Air circulates between approximately 30° and 60° in the Ferrel cells. Near the surface of the earth, the air flows toward the pole and eastward; high in the troposphere the air circulates back toward the equator and toward the west. Beyond the Ferrel cells are the Polar cells. Air circulates within these cells between approximately 60° N and the North Pole, blowing in an easterly direction, or between 60° S and the South Pole [59].

Near the surface of the earth, north–south wind flow can be of the same magnitude as east–west flow; in the middle and upper troposphere, west to east flow predominates. Wind speeds in the troposphere generally increase with altitude and are typically strongest in the winter months. The net effect is that winds in the middle-latitude troposphere generally blow from the west to east, and consequently most intercontinental pollutant transport in the troposphere is from west to east. Short-term weather systems such as cyclones and thunderstorms can disrupt these near-earth wind patterns and affect pollution transport [60].

Jet streams of fast-moving air ride along the tropopause. Although these winds typically blow at 129 to 225 kilometers per hour (80 to 140 miles per hour), they can reach speeds of more than 443 kilometers per hour (275 miles per hour). The two primary jet streams in each hemisphere blow to the east: the Polar jet stream located near latitude 45°, and the Subtropical jet stream located near 30°. The location and speed of these winds vary seasonally [60,61].

Two primary mechanisms account for pollutant transport. The first tends to act on a regional scale; the second, global.

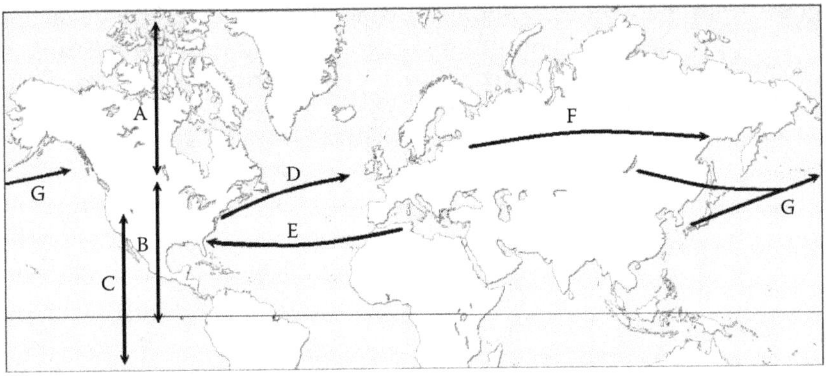

FIGURE 2.4
Major atmospheric transport pathways and transport timescales. A) Midlatitudes–Arctic exchange (1–4 weeks). B) Midlatitudes–Tropics exchange (1–2 months). C) Northern Hemisphere–Southern Hemisphere exchange (approx. 1 year). D) North America to Western Europe (3–13 days). E) Northern Africa to North America (1-2 weeks). F) Eastern Europe to Asia (1–2 weeks). G) Eastern Asia to North America (4–7 days). (Reprinted from the National Academies Press, Copyright 2009, National Academy of Sciences. With permission.)

In the first mechanism, air pollutants emitted to the planetary boundary layer (PBL) can mix into background conditions and move by advective transport (i.e., with the wind) throughout a region. The degree of mixing depends on the turbulence that results from temperature gradients [60]. Air pollutants in the PBL may rise higher into the troposphere or may be carried back to the surface of the earth with temperature-driven air flow. Some pollutants may sorb to solids; precipitation may return those solids or dissolved contaminants back to the ground. Chemical and photochemical reactions may degrade some substances.

The second pollutant transport mechanism can act more globally. Wind currents can lift localized pollutants up out of the PBL and into the free troposphere. The meteorological conditions off the east coasts of North America and Asia, in particular, can drive this form of transport. Once in the free troposphere, strong winds such as the jet streams can move pollutants for long distances without significant dilution or removal [60].

The net effect is this [60, p. 1]: "Any air pollutant with an atmospheric lifetime of at least three to four days may be transported across most of a continent, a week or two may get it across the ocean, a month or two can send it around the hemisphere, and a year or two may deliver it anywhere on Earth."

Figure 2.4 [60] shows how atmospheric transport pathways can carry airborne pollutants around the globe. Such global transport may not occur in a single step or stage. Scientists have coined the terms "global distillation" and the whimsically named "grasshopper effect" to describe a hypothesis that explains how the long-range transport of air pollutants may happen in stages. Global distillation, as the name implies, refers to the sequential volatilization and condensation of a substance. A pollutant released to the atmosphere migrates with the winds until it meets cooler temperatures that change the state of the pollutant and it returns to earth to sorb to soil, overwinter in snow, or dissolve in surface water. When the temperature rises again, perhaps during the next summer season, the compound vaporizes again and can migrate with the prevailing winds. A series of distillation steps can gradually carry a pollutant far from its point of origin, "hopping" across the globe until it reaches steady state in the cold regions of the Arctic or Antarctic. Just as in a distillation column or chromatographic column, this process can effectively separate (or fractionate) substances according to their physicochemical characteristics. More volatile compounds may travel farther than their less-volatile cousins. Persistent semivolatile substances, depending on environmental conditions such as temperature, precipitation, bioturbation, and the organic content of the soils with which they come into contact, are of particular concern with respect to the potential for long-range transport [62].

Although some evidence supports the global distillation hypothesis, other scientists have proposed competing explanations for global transport. The differential removal hypothesis, for example, focuses on the fact that even persistent compounds have some potential to degrade. The originators of this hypothesis postulated that fractionation results not from temperature-driven "distillation," but from different loss rates from the atmosphere that manifest along a gradient of remoteness from emission sources [63].

In sum, the precise mechanism for fractionation of chemical substances during long-range transport is not yet clear. However, scientists have amassed data which show that such transport occurs. We look at that transport for one persistent compound, HCB, in a detailed case study in Chapter 4.

2.1.3 Modeling Fate and Transport of Chemicals in the Environment

Numerous models are available to simulate the fate and transport of chemicals in the environment. These models reflect the multitude of processes described above to varying degrees of complexity and accuracy. They also represent different distance scales, from local to regional to global.

Multimedia environmental models often incorporate the concept of fugacity into mass balance calculations. As pioneered by Dr. Donald Mackay and described in Table 2.3, fugacity models can reflect four levels of sophistication [64]. Level III fugacity models are commonly used to describe the fate and transport of a chemical released to the environment that undergoes degradation and advective transport between compartments. One such model is described below.

The US Environmental Protection Agency's (EPA's) EPISuite® software, for example, contains a Level III fugacity model based on Dr. Mackay's EQC model [65]. EPISuite allows the user to estimate how a chemical partitions between compartments and its overall persistence in the environment. The model represents four main compartments: air, water, sediment, and soil. The software essentially solves a series of equations that represent advection

TABLE 2.3
Multimedia Fugacity Models, Levels I–IV

Level	Premise	Application
I	Chemical is conserved; reaches equilibrium between different phases.	Indicates where a chemical is likely to partition.
II	System at interphase equilibrium; chemical concentrations reduced by advection and degradation.	Provides first approximation of chemical persistence.
III	System is not at interphase equilibrium, that is, substances can transfer between compartments (steady state assumed); chemical concentrations reduced by advection and degradation.	Basis for some screening models, such as those used in US EPA's EPIW software.
IV	Adds to the complexity of Level III model by describing the time-dependent or dynamic behavior of chemicals.	More complex applications.

between air, water, and soil, and the first-order biodegradation of the substance at steady state. These same processes are generally embedded in other models, fugacity-based or not, as they are the fundamental mechanisms that determine the fate and transport of many chemicals in the environment.

Although most multimedia environmental models generally reflect the same fate and transport mechanisms, they do so at different scales. Some models, such as EPISuite, describe a "generic" environment. Other models are specific to certain regions or else simulate the global transport of chemical substances [66].

We return to some of the models used in exposure assessment in Chapter 3. In the meantime, let's look at one illustrative application of a multimedia environmental model to a real-world scenario on a relatively small scale.

Many readers have probably used products containing 1,4-DCB. Such products have been sold in the European Union and United States for mothproofing woolen clothing and as deodorizers, for example, as a "toilet block," or a solid block hanging from the rim of a toilet bowl. In 1980, the US EPA identified 1,4-DCB as a likely contaminant in domestic sewage as a result of its use in insecticides sprayed on fruit, household cleaners, bathroom deodorants, toilet bowl cleaners, spray household deodorants, diaper cleaners, fabric dyes, and upholstery and rug cleaners [67–69]. A less than glamorous starting point for life cycle assessment concepts, to be sure, but 1,4-DCB provides an interesting example. It illustrates the unintended consequences of consumers' choices; in its use as a toilet block it has a nearly direct route to the environment, and it has been the subject of intense regulatory scrutiny in the European Union (as discussed further in the next chapter) and other geographies.

A mass balance on 1,4-DCB could encompass all of the releases to the environment from the point of manufacture of 1,4-DCB to the disposal of a used product containing the substance. To illustrate the tenets of mass

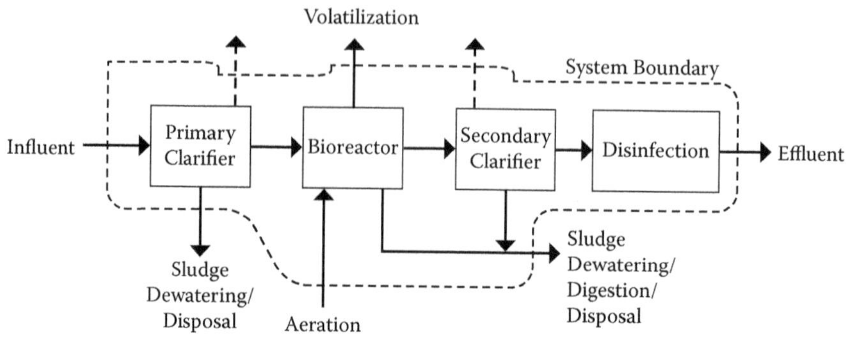

FIGURE 2.5
Simplified process flow diagram for a wastewater treatment plant.

balance and environmental modeling, let's define the system for the mass balance as one portion of this life cycle: the wastewater treatment plant receiving sewage discharges containing the residues of 1,4-DCB from the use of toilet blocks and other consumer products. The system boundary encompasses all of the unit operations within the treatment plant. The next step is to assess the input, output, generation, and consumption of the substance within the system.

Figure 2.5 illustrates a simplified version of that system. (The schematic omits certain treatment units, such as the initial removal of large solids, for simplicity. Furthermore, it does not include disinfection or tertiary treatment of wastewater. Both of those unit operations could potentially degrade an organic chemical or otherwise remove it from the effluent.)

1. *Influent:* Wastewater entering the treatment plant contains 1,4-DCB from its use in toilet blocks. Sewage from buildings that use these toilet blocks or other sources of 1,4-DCB is diluted by sewage from those which do not. Variables include flow, variation in flow with time, and concentration.
2. *Primary clarifier:* This settling tank removes gross solids from sewage by virtue of the difference between the density of the solids and the density of wastewater. Substances such as 1,4-DCB may volatilize, sorb to solids, biodegrade, or flow with the wastewater to the next tank. Solids removed from the tank are dewatered and removed for disposal. Critical variables include the retention time in the tank and the tank dimensions, which can affect the degree of solids removal, and the volatility, tendency to sorb to solids, and degradability of the compound itself.
3. *Bioreactor:* This is also referred to as secondary treatment. In a typical unit, a suspension of activated sludge (essentially, a consortium of microorganisms) is aerated and mixed to enhance the biodegradation

of organic substances. Volatilization and sorption to solids may also remove organic substances from the wastewater. Important process variables include the retention time, aeration rate, and amount of sludge recycled. The properties of the substance, including volatility, tendency to sorb to solids, and degradability also determine its fate.

4. *Secondary clarifier:* Wastewater containing activated sludge flows from the bioreactor to the secondary clarifier, which removes solids from the wastewater stream. A fraction of the solids are recycled to the bioreactor; the remainder is removed from the system for dewatering. Waste sludge may undergo additional treatment, including anaerobic digestion (to further reduce the volume and organic content) and disinfection. The residuals may be landfilled, incinerated, or used to make fertilizer.
5. *Disinfection:* Treated wastewater is disinfected before discharge to the receiving water.

As even this brief summary shows, evaluating the fate of a chemical substance during its life cycle requires that the assessor consider many steps that can affect the fate and transport of a chemical substance. Each step, captured in one or more mathematical relationships, requires the modeler to make numerous assumptions to simulate the range of real-world conditions.

Product stewards often turn to existing software to perform a mass balance like our hypothetical example of 1,4-DCB entering a sewage treatment plant. One such model is the Sewage Treatment Plant (STP) Model [70]. This Level III fugacity model simulates the removal of a substance by volatilization, sorption to solids, and biodegradation based upon its solubility, vapor pressure, octanol–water partition coefficient (which gives one measure of the tendency to sorb to solids), and biodegradability or half-life. It reflects primary and secondary treatment as illustrated in Figure 2.5. It does not include disinfection or tertiary treatment of wastewater, nor does it model the fate of a substance when the primary or secondary sludge undergoes further treatment or disposal.

The first question in applying the model is the influent concentration. As a first approximation, a concentration of 0.01 milligrams per liter (mg/L) was input to the STP model. This concentration is not based on any specific current data regarding the level of 1,4-DCB in domestic sewage. It is the model default value. As a starting point it is not unreasonable: a 1978 US EPA study of priority pollutants in the influent to 40 publicly owned treatment works, which reflected domestic and industrial contributions, detected 1,4-DCB in 17% of the samples at concentrations up to 0.200 mg/L [71].

The software performs the mass balance using built-in default values for the properties of 1,4-DCB and the configuration and operation of a "typical" wastewater treatment plant. It predicts that treatment will remove a total of 72.6% of the mass of 1,4-DCB that enters the plant, as follows:

- *Primary clarifier:* 9.55% removal (primarily by sorption to sludge, which accounts for 5.47%)
- *Aeration tank:* 57.6% removal (primarily by biodegradation, which accounts for 39.9%)
- *Secondary clarifier:* 5.37% removal (primarily by biodegradation, which accounts for 2.93%)

The effluent in this hypothetical example would contain 27.4% of the influent concentration, or 0.00279 mg/L DCB. That concentration could be compared to toxicity data to assess the potential risk from exposure but for one important caveat: it was based on an assumed influent concentration and calculated using many assumptions and thus is unlikely to represent the actual conditions at any specific environmental location.

Different models or model variants produce different values. The US EPA's EPI SuiteTM v4.11 [72] software contains a module, STPWin, that uses fugacity modeling to approximate the fate of an organic compound entering a model sewage treatment plant operating at 25°C. STPWin is a version of the STP model. It produces different estimates of the degradation of 1,4-DCB depending on assumptions regarding the rate of biodegradation, or half-life. Allowing the model to calculate the degradation rate with the BioWin module produces a removal estimate of 63.8% removal (primarily due to two processes: 20.48% by biodegradation and 35.07% by volatilization).

As this brief example shows, one can readily produce different estimates of the fate and degradation of a substance by using different models or model variants and input factors. The calculations summarized above, which approximate the fate of 1,4-DCB using the default values in two widely used model variants, produced removal estimates of 64 and 73%. More interesting than the difference in the total removal is the predicted fate of the 1,4-DCB removed from the wastewater. The STP model estimated that approximately 45% of the influent 1,4-DCB would biodegrade and 20% volatilize to air; STPWin with the Biowin module suggested that approximately 21% would biodegrade and 35% volatilize.

The results of bench- and pilot-scale studies on the biodegradation of 1,4-DCB in sewage treatment provide a reference point for these model results. These studies have shown up to 100% removal of 1,4-DCB. Volatilization removed up to 68% of the mass and biodegradation accounted for much of the remainder. Little sorbed onto sludge solids. Elimination of 1,4-DCB in full-scale sewage treatment plants has been reported as 60 to 74% [73].

This apparent difference in modeled projections of the primary mode of removal is significant with respect to the life cycle of the chemical and the potential for exposure. It illustrates two key points: why models should not be used blindly, and that the models simulate processes that may themselves vary widely in their application.

We return to 1,4-DCB and to the use of multimedia environmental models in Chapters 3 and 4. As this relatively simple discussion shows, life cycle assessment reflects a tremendous paradox. The principle of conservation of mass is simply stated and readily understood. But applying that principle to the manufacture and use of a chemical to anticipate environmental consequences is a tremendously complex undertaking that requires myriad input data—with all their variability and imprecision—and numerous assumptions regarding real-world conditions. Even rigorous models can only approximate the complexity of chemical behavior and environmental interactions, which are complicated many-fold by the regional variations in product usage, waste management, and environmental conditions. Nonetheless, those models can produce useful first approximations of the concentrations of chemicals in the environment to assess the consequences of their use.

2.1.4 Recapitulation

- Matter is neither created nor destroyed; one can quantify the fate and transport of a chemical substance in the environment by a mass balance that accounts for phase changes, convective transport, biological reactions and interactions, and abiotic reactions.
- Of particular focus in many environmental assessments and regulatory schemes are chemicals characterized as persistent and bioaccumulative. Persistence refers primarily to resistance to biodegradation, which may result from a molecular structure that is hydrophobic or contains strong chemical bonds. Persistence is sometimes defined by a half-life in water greater than two months or a half-life in sediment or soil greater than six months. Bioaccumulation, which refers to the accumulation of a chemical substance within a living organism, may be of concern if testing indicates that the substance has a BAF in aquatic species greater than 5,000 or a log K_{ow} greater than five. (These benchmark values can vary by regulatory scheme.)
- Advection with wind or water can transport a chemical substance locally or far from its point of release if the substance resists degradation. A half-life in air of greater than two days or a half-life in water on the order of 8 to 130 days may signal the potential for long-range transport.
- Multimedia environmental models perform mass balances on chemicals in the environment and can predict the resulting concentrations in various environmental compartments. Many models incorporate the concept of fugacity to allow for the deviations of the natural world from ideal behavior.

- Every mass balance on the fate and transport of a chemical in the environment requires numerous assumptions; the results should not be used blindly or assigned too much weight as absolute predictors of environmental concentrations.

2.2 Characterizing Risks

Models that predict the concentrations of chemicals in the environment and efforts to regulate chemical manufacture and use inevitably provoke the question of what concentrations are acceptable. Some answer that question simply with the conviction that no amount of anthropogenic chemicals in the environment or our bodies is acceptable. But many people and most regulatory programs answer that question based on an assessment of the estimated risks to human health and the environment from exposure to anthropogenic chemicals. This discussion briefly describes the common paradigm for assessing human health and environmental risks in order to provide context for subsequent chapters regarding regulations on chemical manufacture and use and case studies.

Risk assessment is a science-based effort to predict the effect of exposure on human health or the environment [74]. It generally reflects a multistep process [75–77]. In the first two steps, the analyst assesses the potential hazards of exposure and estimates the dose–response. Next, the potential exposures are described and estimated. Finally, the analyst combines the estimates of hazard and exposure to quantify the potential risks.

Each of these steps combines numerous assumptions that allow the risk assessor to extrapolate from the available data to potential real-world scenarios. A detailed description of the art and science of risk characterization is beyond the scope of this book; the reader should simply be aware that the process can reflect both scientific thinking and subjective judgment [78], and that in consequence risk assessments can produce a variety of estimates of risk from the same data.

Risk assessment differs from risk management. Risk management is a "decision-making process involving considerations of political, social, economic, and technical factors with relevant risk assessment information relating to a hazard so as to develop, analyse, and compare regulatory and non-regulatory options and to select and implement appropriate regulatory response to that hazard" [74, p. 10]. In short, it reflects a determination of the level of acceptable risk and the appropriate measures to achieve that goal. Chapter 3 describes how regulations and voluntary efforts to control the manufacture and use of chemicals reflect both risk assessment and risk-management decisions.

2.2.1 Hazard Identification and Characterization

Hazard assessment begins with assembling information regarding the potential physical/chemical hazards or toxicity of a chemical. The risk assessor may seek to characterize:

- Physical/chemical hazards, such as flammability or explosivity
- Ecotoxicity, particularly to three aquatic trophic levels: algae, aquatic invertebrates, and fish, and sometimes including terrestrial organisms
- Toxicity to humans, typically based on extrapolations from effects on animals allowing for one or more uncertainty factors

Hazard characterization may consider both acute and chronic effects. Acute effects occur in the short term; chronic effects result from longer-term exposure.

This information may derive from laboratory testing. Alternatively, toxicologists may predict hazards from a structure–activity relationship (SAR), which is "the relationship of the molecular structure of a chemical with a physicochemical property, environmental fate attribute, and/or specific effect on human health or an environmental species. These correlations may be qualitative (simple SAR) or quantitative (quantitative SAR, or QSAR)" [79, p. 1]. Both "read across" from the behavior of one chemical to that of a similar chemical and the practice of assessing chemicals by categories are forms of SAR [80,81].

Assessment endpoints have often included those listed in Table 2.4; the endpoints considered in a risk assessment may depend on the nature of the substance and the applicable regulatory regime, among other factors, and can include effects other than those tabulated (e.g., see [82]). Table 2.4 also includes environmental behaviors relevant to short- and long-term exposure that may be considered during the risk assessment.

Toxicologists and risk assessors pay particular attention to chemicals that are persistent, bioaccumulative, and toxic (PBT), or very persistent and very bioaccumulative (vPvB). Persistent chemicals do not readily degrade and can linger in the environment for many years; bioaccumulative substances can significantly bioconcentrate in the food chain; and certain substances may be designated as particularly toxic if they are carcinogens, mutagens, or reproductive toxicants (CMR) or based on a comparison of the dose–response to thresholds specified by regulatory agencies (e.g., [83]).

Toxicologists typically quantify the response of test organisms to a chemical dose and then extrapolate from those test data, incorporating various margins of safety, to develop benchmarks used in risk assessment. Some of the most important benchmarks are described below using the jargon applied to the process in the European Union and the United States [84–86].

The derived no-effect level, or DNEL, is the level of exposure to a substance above which humans should not be exposed. Risk assessors use

TABLE 2.4
Assessment Endpoints

Concern	Short Term/Acute	Longer Term/Chronic
Physical Hazards	Flammability Explosivity Corrosivity	—
Environmental Behavior	Solubility Volatility Sorption Oxidation-reduction potential Biodegradability Hydrolysis	
Ecotoxicity	Acute toxicity—fish Acute toxicity—aquatic invertebrates (daphnia) Growth inhibition—green algae	Bioaccumulation Chronic toxicity—fish Chronic toxicity—aquatic invertebrates
Toxicity (Human Health)	Acute toxicity—oral Acute toxicity—inhalation Acute toxicity—dermal	Carcinogenicity Mutagenicity Reproductive toxicity Developmental toxicity Neurotoxicity Systemic toxicity Irritability Sensitization

this benchmark to characterize the toxicity of chemicals believed to have a "threshold effect"; that is, this benchmark reflects the implicit assmption that at a very low dose (i.e., below the DNEL) negligible harm would occur from exposure. A DNEL is specific to a route of exposure (oral, dermal, or inhalation), type of exposure (worker or general public), duration of exposure, and effect. The units are specific to the route of exposure: milligram per cubic meter (mg/m^3) for inhalation, and milligrams per kilogram body weight per day (mg/kg/d) for oral and dermal exposure. The term DNEL is used in the European Union, however, the US EPA uses the terms oral reference dose (RfD) and inhalation reference concentrations (RfC).

Toxicologists generally posit that no threshold exists for some effects, notably for most carcinogens. Such effects are characterized by a derived minimal effect level (DMEL) in the European Union; in the parlance used in the United States, risks are characterized using oral slope factors and oral or inhalation unit risks. A DMEL represents the exposure level that corresponds to a specified level of risk to the exposed population. In the case of a chemical known or suspected to be a human carcinogen, this exposure level corresponds to a risk of one excess case of cancer in an exposed population. For a hypothetically exposed population of one million people, for example, this risk level of one excess case of cancer in the population of one million is abbreviated

10^{-6}. The cancer risk decision points used for lifetime exposure of the general population are generally in the range of 10^{-4} to 10^{-6}, depending on the regulatory program. DMELs are typically presented in units of mg/kg/d.

Risk assessors also work with benchmarks that characterize the potential risk to the environment. The predicted no effect concentration (PNEC), in the terminology used in the European Union, is the concentration of a substance below which adverse effects will most likely not occur during long-term or short-term exposure. PNECs may be derived for aquatic organisms, terrestrial organisms, or the bacteria in a sewage treatment plant, for example, and are typically presented in units of mass per volume or mass. The US EPA has utilized similar parameters, the maximum acceptable toxicant concentration (MATC), and, as discussed further in Chapter 3, the concern concentration.

The preceding discussion of hazard characterization noted that dose–response factors go by different names in different regions. More important, the actual values of the dose–response factors for a single chemical can vary under different regulatory regimes. Consider the findings of one study that compared the long-term systemic inhalation DNELs for the general population, derived under REACH in the European Union with the inhalation RfCs developed by the US EPA [87]. The two values varied greatly for a single chemical. For example, the inhalation RfC for 1,1-dichloroethylene is 0.2 mg/m^3 and the inhalation DNEL for long-term systemic effects is 3.5 times higher, at 0.7 mg/m^3. For trichloroethylene, the inhalation RfC is 0.002 mg/m^3; the corresponding DNEL is 13.7 mg/m^3, a factor of 6,850 higher. The authors of the study identified four possible reasons that might account for such differences:

- DNELs may reflect data available under REACH that were not available to the US EPA.
- Some of the default assumptions utilized to derive DNELs and RfCs differ.
- DNELs are derived by manufacturers/importers and RfCs by US EPA staff.
- Under REACH, there is no formal peer-review process for DNELs before submittal to the European Chemical Agency (ECHA); RfCs, in contrast, undergo extensive peer review and a public comment period.

Regardless of the cause, the differences in interpretation of toxicological data illustrated here can obviously affect both the outcome of a risk assessment and the risk management decisions based on such an assessment.

2.2.2 Exposure Assessment

Exposure assessment is the "evaluation of the exposure of an organism, system or (sub) population to an agent (and its derivatives)" [74]. An exposure assessment may be as simple as collecting representative air monitoring data in a manufacturing facility. More commonly, risk assessors utilize a mass balance to estimate the concentration of a substance in air, water, soil, sediment, or sewage, and contemplate the exposures that could result from those concentrations.

Many regulations and regulatory guidance specify the format and content of exposure assessments. All generally share the same sequence of steps [88]. The first step in the process is to develop one or more exposure scenarios. This amounts to telling a story about how the exposure might occur. The scenario begins, as does any mass balance, by defining the system within which exposure occurs. An exposure assessment may be performed on a local, regional, national, or continental scale. The assessor also defines the timescale for the exposure, short or longer term, depending in part on the persistence of the chemical.

The next step is to identify the source(s) of a chemical and the nature of the release(s) of that chemical, including the amount and to what media. The story continues with a description of how the chemical moves through the environment, considering advection and transport between phases, the reactions it undergoes, and the potential to bioaccumulate. Finally, the exposure scenario describes who or what is exposed (i.e., the receptor, ecosystem, or habitat).

The assessor combines the assumptions stated in the exposure scenario with information about the behavior of the chemical to perform a mass balance that produces an estimated exposure concentration, known as a predicted environmental concentration (PEC), in each of the relevant compartments.

Exposure assessment is often an iterative process that begins with numerous simplifying assumptions. For example, the assessor might initially assume that no degradation or advective transport occurs, or that all of the substance is bioavailable. Such assumptions produce a worst-case estimate of exposure. If that estimate would not present a significant risk, then no further exposure assessment may need to be done. Alternatively, the assessor may choose to refine his assumptions, perhaps by collecting additional data to quantify more accurately some aspect of the compound's fate and transport, or by incorporating a probability distribution for the values of critical parameters (e.g., Monte Carlo simulation). He may also choose a more sophisticated model to perform the mass balance.

2.2.3 Risk Characterization

Risk characterization combines the results of the hazard assessment and exposure assessment to project the potential risk to human health or the environment. The way in which it is done in the European Union under REACH illustrates the process [76].

A risk characterization comprises a series of evaluations, each pertaining to a specific population, route of exposure, and frequency and duration of exposure. (ECHA refers to this combination of factors as the "exposure pattern".) Assessors characterize the risk to human health by comparing the estimated exposure level for a given exposure pattern with the lowest DNEL/DMEL value for that exposure pattern. If the exposure could exceed the DNEL/DMEL, resulting in a risk characterization ratio (RCR) above 1, then the risk could be significant.

Similarly, an environmental risk characterization is carried out by comparing the predicted exposure concentrations (PECs) with the corresponding PNECs. If a PEC exceeds the corresponding PNEC, then the risk may be significant. A series of comparisons may be performed to assess the risks at different trophic levels, in different media, or at different scales (e.g., regional versus local).

The process is often iterative. If an assessment made using worst-case assumptions shows that the risk would be acceptable, then no refinement is necessary; if the worst-case calculations show that the risk would be significant, however, then the assessor often gathers additional information to refine the assumptions.

Even the most sophisticated risk assessment has limitations. It involves numerous assumptions about both exposure and hazard. Exposure assessments typically reflect modeled concentrations or extrapolations from measured data. The degree of exposure by different individuals may vary, and their response can depend on factors such as general health, genetic predisposition, or other factors. Dose–response factors are typically extrapolated from animal studies and thus inherently introduce the uncertainty of relating the response of laboratory animals to that of humans or one of the many species in an ecosystem. The endpoints characterized may not include all of the potential effects; for example, the potential for endocrine disruption has not been considered in many risk assessments and in fact standardized testing methods were not published until approximately 2007 or later [90]. And risk assessment tools only model relatively simple scenarios. They rarely account for exposure to multiple chemicals, or fully account for the effects on a complex web of organisms in an ecosystem.

Despite these limitations, risk characterization is the tool available to us to put chemical exposures into perspective. It often comes into play in the life cycle assessment of a chemical, whether that assessment is a voluntary effort or driven by a regulatory requirement.

2.2.4 Recapitulation

- Risk assessment usually follows a four-step paradigm: hazard identification, hazard characterization, exposure assessment, and risk characterization. (Some regulatory risk assessment protocols conflate the first two steps.) Each step reflects numerous assumptions. Consequently, the process is often iterative and may include a sensitivity analysis. The four steps are as follows:

- *Hazard identification:* Evaluation of whether a chemical substance has the potential to cause harm in the short term (acute effects) or long term (chronic effects).
- *Hazard characterization or dose–response assessment:* Examination of the quantitative relationship between exposure and effects.
- *Exposure assessment:* Development of qualitative scenarios describing potential exposures and quantification of predicted frequency, timing, and concentrations of exposure.
- *Risk characterization:* Comparison of exposure levels to acceptable levels to determine whether potential risk is tolerable. In a human health risk assessment (using the terminology common in the European Union), the estimated exposure level is compared to the lowest DNEL/DMEL value for that exposure pattern to determine the RCR; in an ecological risk assessment, the PEC is compared to the PNEC.

- Risk management differs from risk assessment; it refers to the choices made in consideration of the potential risks and technical and socioeconomic factors that influence the risks and benefits.

2.3 Life Cycle Assessment

What is life cycle assessment and how does it bear on the central questions of this book? Let's begin with some definitions, in order to be clear about the matters at hand.

Resource and environmental profile analysis (REPA), the forerunner to the current practice of life cycle assessment, focused on quantifying the energy requirements and emissions of a product or process but not the impacts on human health or the ecosystem. Ideally, according to the originators of REPA, the analysis would be linked to a risk assessment of emissions related to a process or product [91,92]. It is worth noting, with respect to the theme of this book, that REPA originated in 1969.* Environmental life cycle considerations did not formally enter into product development or modification before that time.

Life cycle assessment (LCA) looks at the potential effects on the environment as a result of the extraction of resources, transportation, production,

* The first REPA attempted to quantify the energy, material, and environmental consequences of the entire life cycle of a Coca-Cola beverage container from the extraction of raw materials to disposal; under consideration were refillable bottles, plastic bottles, and cans.

use, recycling, and discarding a product [93]. It is, in short, a "compilation and evaluation of the inputs, outputs and the potential environmental impacts of a product system throughout its life cycle"[2, p. 2]. Life cycle impact assessment (LCIA) is one critical phase of an LCA. It is "aimed at understanding and evaluating the magnitude and significance of the potential environmental impacts for a product system throughout the life cycle of the product" [2]. The term LCA originated at a conference of the Society of Environmental Toxicology and Chemistry (SETAC) in 1990; practitioners have since codified LCA in formal guidelines, notably *ISO 14040: Environmental management—Life cycle assessment—Principles and framework* [2] and *ISO 14044: 2006—Environmental management—Life cycle assessment—Requirements and guidelines* [94].

Life cycle thinking (LCT) is a holistic approach to the design or optimization of a product or process. In the words of one authority [95], LCT:

> ... seeks to identify possible improvements to goods and services in the form of lower environmental impacts and reduced use of resources across all life cycle stages. This begins with raw material extraction and conversion, then manufacture and distribution, through to use and/or consumption. It ends with re-use, recycling of materials, energy recovery and ultimate disposal.

Life cycle thinking is sometimes referred to as life cycle perspective [96]. The primary objective of such thinking or perspective is to avoid shifting the burden from an action by minimizing impacts at one stage of the product's life cycle or in one geographic region only to increase the impact in another place or time.

Life cycle sustainability assessment (LCSA) adds two elements to an environmental LCA: social and economic factors. The term refers to "the evaluation of all environmental, social and economic negative impacts and benefits in decision-making processes towards more sustainable products throughout their life cycle" [93].

These various practices intersect at the fundamental concept of considering the human health and environmental consequences of making, using, and discarding a product.

2.3.1 Basic Tenets

Simply put, life cycle assessment is [97, p. 2]:

> ... a technique to assess the environmental aspects and potential impacts associated *with a product, process, or service, by:*
>
> - Compiling an inventory of relevant energy and material inputs and environmental releases
> - Evaluating the potential environmental impacts associated with identified inputs and releases

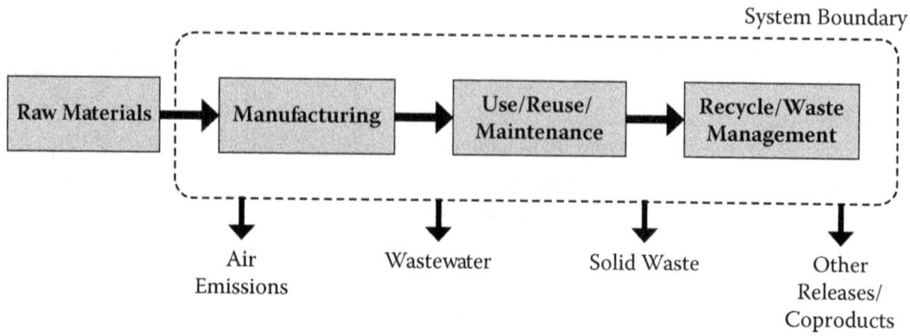

FIGURE 2.6
Life cycle assessment.

- Interpreting the results to help decision-makers make a more informed decision.

A formal LCA comprises four steps: goal definition and scoping, inventory analysis, impact assessment, and interpretation. Figure 2.6 illustrates the process with respect to a material or mass balance, omitting the components of energy balance. Each step is described briefly below [2,93,97,98] as it would be applied to a product or process rather than a service, and with a focus on mass inputs and outputs rather than energy, noise, or other considerations sometimes taken into account in LCA.

2.3.2 Goal and Scope Definition

The first step is to define the goals and scope of the LCA. The overarching goal of an LCA is to choose the best product or process with the least effect on human health and the environment. The assessor may also have other, secondary goals for the work related to business or policy objectives or to the audience for the results (e.g., whether the work will inform decision makers in a corporation or is intended for other stakeholders). The scope must describe the work to be done to meet these goals. It specifies, among other things:

- The product or process under analysis and the function delivered.
- The system boundary for the mass balance.
- The functional unit for the analysis. The functional unit is analogous to the basis for a mass balance. It quantifies the function of the process or product in a way that allows the analyst to normalize the data for different alternatives. For example, the functional unit for an LCA comparing a liquid detergent to a powdered detergent might be the amount of each product required for a certain number of loads of laundry.

- The impact categories to be considered, typically based on potential impacts on human health, ecological health, and resource depletion, and methodology of impact assessment.
- Anticipated data needs and data quality objectives.
- Assumptions and limitations.

2.3.3 Life Cycle Inventory Analysis

Life cycle inventory (LCI) analysis is the second step in an LCA. The assessor typically diagrams the steps (or unit processes) in the product or process life cycle as the basis for this analysis. For example, Figure 2.5, which showed the unit operations in the treatment of wastewater, diagrams one portion of the life cycle of 1,4-DCB.

The assessor then quantifies the inputs and outputs for each step (or unit process) in the product or process life cycle. Inputs may derive from direct process knowledge, literature values, databases maintained by industry organizations or governmental agencies, or measurements. Outputs are estimated using mass balance calculations, noting uncertainties and assumptions. These calculations quickly become quite complicated due to the number of steps in the life cycle and the inputs and outputs that occur at different times and places. As a result, LCA commonly requires software tools [99,100]. Further, it may be worthwhile to perform a sensitivity analysis at the conclusion of this phase and reiterate the calculations with a refined set of assumptions.

2.3.4 Life Cycle Impact Assessment

The third step in a LCA is impact assessment. The life cycle impact assessment (LCIA) is [97, p. 46]:

> ... the evaluation of potential human health and environmental impacts of the environmental resources and releases identified in the LCI. Impact assessment should address ecological and human health effects.... A life cycle impact assessment attempts to establish a linkage between the product or process and its potential environmental impacts.

LCIA includes three mandatory elements:

1. Selection of impact categories, category indicators, and characterization models. Impact categories are environmental issues of concern. Some, such as ecotoxicity (terrestrial or aquatic) and human health toxicity, are familiar from the discussion of risk characterization in Section 2.2. Others do not typically enter into environmental risk assessments outside the context of LCA. Such impact

categories include climate change, resouce depletion, land use, or water use. A category indicator is a quantifiable representation of an impact category.

2. Classification, which is the assigment of LCI results to the selected impact categories. Consider, for example, the case where the LCI comprised a multimedia enviromental model to predict the distribution of a chemical in soil, sediment, air, or water. The masses (or concentrations) in those different media would then be "assigned" to the impact categories of ecotoxicity (terrestrial or aquatic) and human health toxicity depending on the likely exposure to the chemical in each medium.
3. Characterization, or calculation of category indicator results. In this step, one converts the LCI results to common units using characterization or equivalency factors to normalize the results, and aggregates the results within an impact category. In an LCA that included climate change as an impact factor, for example, all greenhouse gases would be expressed in terms of carbon dioxide (CO_2) equivalents by multiplying the LCI results by a scientifically based CO_2 characterization factor and then summing the resulting impact indicators to provide an overall indicator of the potential environmental effect. Similarly, for the impact category of aquatic ecotoxicity, one might normalize the estimated concentration or mass of several different chemicals to the equivalent concentration or mass of a representative chemical based upon the relative values of a toxic effects concentration such as LC50 (i.e., lethal concentration to 50% of the organisms tested under the specified test conditions).

LCIA can also include three additional steps:

4. Normalization, or calculating the magnitude of category indicator results relative to reference information. One might, for example, normalize the results to represent the impact caused in one year per capita. This step can allow the assessor to better understand, interpret, and communicate the results of the analysis.
5. Grouping or aggregation. In this step, the assessor assigns impact categories into one or more sets to aid in interpretation or ranking of the LCA results. These sets may reflect scale (local, regional, global) or priority (high, medium, low, based on value judgments), for example.
6. Weighting, also described as valuation. In this step of an LCIA, the assessor assigns weights or relative values to the different impact categories based on their perceived importance or relevance. Obviously subjective, this step enables the process to mirror stakeholders' values.

Finally, and particularly important due to the numerous assumptions and value judgments in the process, the results must be critically reviewed and thoroughly documented.

The preceding description of LCIA alludes to risk characterization steps; however, LCIA differs from risk assessment in several important respects [1]. LCIA seeks to link a system, as defined in the LCA, with potential effects in order to allow for a relative comparison of the potential impacts from alternative products or processes. LCIA is not designed to quantify accurately the risk of actual harm to an exposed population, ecosystem, or resource.

2.3.5 Interpretation

The fourth step in an LCA is the structured interpretation of the results within the context of the original goal and scope of the study. Interpretation includes:

- Identification of significant issues
- Checks on completeness, sensitivity, and consistency of the analysis
- Development of conclusions, limitations, and recommendations

2.3.6 Recapitulation

- LCA is a formalized methodology used to assess the relative impacts of products, systems, and services by compiling and assessing the inputs, outputs, and the potential environmental impacts of a system throughout its life cycle.
- Although LCA contains elements of the tools discussed earlier in this chapter—mass balance, multimedia modeling of environmental fate and transport, and risk characterization—it is a distinct discipline with its own jargon, precepts, and limitations.
- LCA concepts are applied in various contexts. Life cycle thinking or perspective incorporates LCA principles into the design or optimization of a product in order to minimize and balance the possible impacts. Life cycle sustainability assessment adds social and economic factors to LCA. Finally, certain regulations reflect elements of LCA, as we explore in Chapter 3.

2.4 Synthesis

The technical tools described in this chapter give us some ability to predict the consequences of chemical manufacture and use and make risk management decisions. Substances released to the environment during raw material sourcing or the manufacture, use, or disposal of a product move through the environment in ways that can be described with physical laws and chemical reactions. Scientists have standardized tests to measure the factors that determine a chemical's behavior and interaction with living beings, whether those interactions degrade the substance or act in some form on the organism itself. Engineers have developed models that allow us to quantify the fate and transport of chemical substances in the environment using mass balance techniques that account for movement between compartments, degradation, and advective transport. Based on those results, risk assessors working within the context of regulatory programs or voluntary life cycle assessments can gauge the potential harm to human health and the environment of the manufacture, use, or disposal of a product containing a chemical substance. But each step in this chain of analyses entails uncertainties and approximations and consequently risk managers rely on imperfect and sometimes subjective input.

Perhaps the most important aspect of these tools, considering the central questions of this book, is when they became available. Figure 2.7 maps the availability of those tools on a timeline showing the number of chemical substances in existence at that time [101].

Let's start with the number of chemical substances in existence. These values represent the number of chemicals designated by a Chemical Abstracts Service (CAS) number. The index originated in 1965 and reflects the registry's efforts to track back through the chemical literature since the early 1900s to

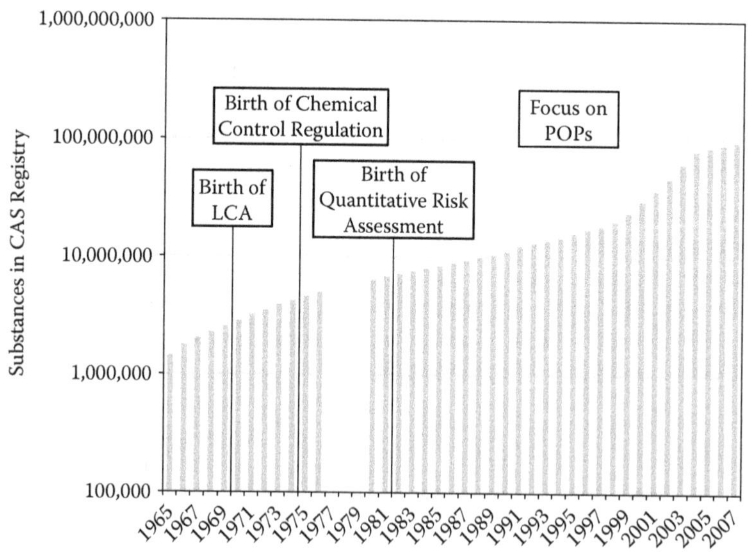

FIGURE 2.7
Availability of tools for assessing life cycle. (**Note:** No data were available for 1977–1979.)

identify individual chemical substances. Newly synthesized chemicals are added to the registry every year. Of the millions of chemicals added to the CAS registry since it originated, most have never left the laboratory bench. In the United States, for example, approximately 80,000 chemicals have been in commerce in the last 40 years and only about 2,000 chemicals account for nearly 95% of chemical production by volume [102]. Thus the numbers of chemicals shown in the figure should not be taken to indicate the number of chemicals in use. They do, however, illustrate the challenge of anticipating the consequences of manufacturing and using chemical substances given the thousands of substances newly created and considered for commercial use each year.

Over a million chemical substances were known when teams of scientists first conceptualized LCA in the late 1960s and early 1970s; that number had increased by nearly tenfold a decade later when regulators began to codify the practices for quantitative chemical risk assessment.

PBT compounds and POPs may present the greatest risks. Alarms regarding POPs in the environment were raised as early as 1962 with Rachel Carson's book *Silent Spring* [103], but not until some 30 years later did scientists begin to understand the global transport mechanisms that could carry such compounds to the most remote parts of the world. By the mid-1990s, when concerns about POPs gained momentum, some 10 million chemicals were on the CAS registry.

Figure 2.7 shows one more factor in addition to these scientific milestones: the rise of chemical control legislation. Early discussions of the need for and scope of such legislation presaged some of the life cycle analysis concepts discussed in this chapter. Notably, in 1971 the US President's Council on Environmental Quality said this [104, pp. 759–760]:

> [It is essential to take] an integrated approach to toxic substances. However, such an approach cannot be accomplished simply by coordinating the activities of existing media-related programs. The activities themselves must be conducted on an integrated basis. Testing to determine the health or environmental effects of a substance must be done in terms of total exposure to the substance, not simply exposure through air or through water. There must exist a capability for integrating the monitoring data from various media and for doing nonmedia analyses, for example, using the materials balance approach. (This approach compares the total amount of a substance produced with the amount appearing in various end uses. A disparity between the two indicates the approximate amount escaping into the general environment.) Finally, there must exist authority to insure that the effects of a new substance are carefully examined before it enters the air, soil, or water.

In Chapter 3, we look at how chemical control legislation in three major economies incorporates and interprets the technical concepts discussed in this chapter.

References

[1] US EPA. 2006. *Life Cycle Assessment: Principles and Practice.* Prepared by Scientific Applications International Corporation for US Environmental Protection Agency, Washington, DC, EPA/600/R-06/060. Pp. 1–2. Available at: http://cfpub.epa.gov/si/si_public_record_Report.cfm?dirEntryID=155087 (accessed November 1, 2013).

[2] International Standards Organisation. 2006. ISO 14040: 2006—*Environmental management— Life cycle assessment—Principles and framework.*

[3] Himmelblau, D.H. 1974. *Basic Principles and Calculations in Chemical Engineering,* third edition, Upper Saddle River, NJ: Prentice-Hall, pp. 73, 148.

[4] OECD. 2000. *OECD GUIDELINE FOR THE TESTING OF CHEMICALS. Adsorption - Desorption Using a Batch Equilibrium Method. Guideline* 106.

[5] Mackay, D. 2001. *Multimedia Environmental Models: The Fugacity Approach,* second edition. Boca Raton, FL: Taylor & Francis/CRC Press, pp. 72–75.

[6] OECD. 2006. *OECD Guidelines for the Testing of Chemicals: Revised Introduction to the OECD Guidelines for Testing of Chemicals,* Section 3. Part 1: Principles and strategies related to the testing of degradation of organic chemicals. Available at: http://www.oecd-ilibrary.org/docserver/download/9730001e.pdf?expires=1388688038&id=id&accname=guest&checksum=3222BAE6FAADBAE19ECFFFCC96B7E34C (accessed January 2, 2014).

[7] US EPA. 2013. *Mercury – Basic Information.* Available at: http://www.epa.gov/hg/about.htm (accessed January 2, 2014).

[8] Scow, K.M. and Johnson, C.R. 1997. Effect of sorption on biodegradation of soil pollutants. *Advances in Agron.* 58: 1–56.

[9] Sei, K., Miyagaki, K., Kakinoki, T., Fukugasako, K., Inoue, D., and Ike, M. 2013. Isolation and characterization of bacterial strains that have high ability to degrade 1,4-dioxane as a sole carbon and energy source. *Biodegradation.* Sep;24(5):665–674. doi: 10.1007/s10532-012-9614-1. Epub 2012 Dec 13.

[10] European Communities. 2002. *European Union Risk Assessment Report: 1,4-Dioxane.* Available at: http://esis.jrc.ec.europa.eu/doc/risk_assessment/REPORT/dioxanereport038.pdf (accessed January 26, 2014).

[11] OECD. 2004. *NEW CHEMICAL ASSESSMENT COMPARISONS AND IMPLICATIONS FOR WORK SHARING.* ENV/JM/MONO(2004)27. Available at: http://search.oecd.org/officialdocuments/displaydocumentpdf/?cote=env/jm/mono(2004)27&doclanguage=en (accessed September 14, 2013).

[12] US EPA. 2012. *User's Guide and Technical Documentation - KABAM Version 1.0.* (K_{ow} (based) Aquatic BioAccumulation Model). Appendix F. Description of Equations Used to Calculate the BCF, BAF, BMF, and BSAF Values. Web page last updated May 09, 2012. Available at: http://www.epa.gov/oppefed1/models/water/kabam/kabam_user_guide_appendix_f.html (accessed January 12, 2014).

[13] US Geological Survey. 2013. *Bioconcentration – Definitions.* Web page last updated July 10, 2013. Available at: http://toxics.usgs.gov/definitions/bioconcentration.html (accessed January 12, 2014).

[14] US Geological Survey. 2013. *Bioaccumulation – Definitions.* Web page last updated July 10, 2013. Available at: http://toxics.usgs.gov/definitions/bioaccumulation.html (accessed January 12, 2014).

[15] US EPA. 2012. *Waste and Cleanup Risk Assessment Glossary – Biomagnification.* Web page last updated December 24, 2012. Available at: http://www.epa.gov/oswer/riskassessment/glossary.htm (accessed January 18, 2014.)

[16] OECD. 2004. *OECD Guidelines for the Testing of Chemicals: Hydrolysis as a Function of pH.* OECD guideline 111.

[17] Stockholm Convention. 2014. *Annex D: INFORMATION REQUIREMENTS AND SCREENING CRITERIA.* Available at: http://chm.pops.int/TheConvention/Overview/TextoftheConvention/tabid/2232/Default.aspx (accessed February 7, 2014).

[18] ECHA. 2012. *Guidance on Information Requirements and Chemical Safety Assessment.* Chapter R.11: PBT Assessment. Version 1.1. Available at: http://echa.europa.eu/documents/10162/13632/information_requirements_r11_en.pdf (accessed January 17, 2014).

[19] Shogren, E. 2014. *The Big Impact of a Little-Known Chemical in W.Va. Spill.* National Public Radio. January 13. Available at: http://www.npr.org/2014/01/13/262185930/mysteries-persist-surrounding-west-virginia-chemical-spill (accessed January 17, 2014).

[20] Berzon, A. and Maher, C. 2014. West Virginia chemical-spill site avoided broad regulatory scrutiny. *The Wall Street Journal.* January 13. Available at: http://online.wsj.com/news/articles/SB1000142405270230381970457931706227356476 (accessed January 17, 2014).

[21] US EPA. 2014. *Toxic Substance Control Act Test Submissions, Version 2.0.* Last Revision: January 18, 2014. http://yosemite.epa.gov/oppts/epatscat8.nsf/ReportSearch?OpenForm (queried January 18, 2014).

[22] Pitter, P. 1976. *Water Res.* 10: 231–235, as cited in: U.S. National Library of Medicine, 2011. Hazardous Substances Data Bank – Methylcyclohexanol. Last Revision Date: 20110503. Available at: http://toxnet.nlm.nih.gov/cgi-bin/sis/search/a?dbs+hsdb:@term+@DOCNO+2910 (accessed January 17, 2014).

[23] Funasaki, N. et al. 1986. *J. Chromatogr.* 361: 33–45. As cited in: Library of Medicine, 2011. *Hazardous Substances Data Bank – Methylcyclohexanol.* Last Revision Date: 20110503. Available at: http://toxnet.nlm.nih.gov/cgi-bin/sis/search/a?dbs+hsdb:@term+@DOCNO+2910 (accessed January 17, 2014).

[24] Howard, P.H., Boethling, R.S., Jarvis, W.F., et al., Eds. 1991. *Handbook of Environmental Degradation Rates.* Chelsea, MI: Lewis, pp. 452–453. As cited in: ATSDR, 2014. Agency for Toxic Substances and Disease Registry (ATSDR) 2013. *Toxicological Profile for Hexachlorobenzene. Draft for Public Comment - June.* Available at: http://www.atsdr.cdc.gov/toxprofiles/tp.asp?id=627&tid=115 (accessed January 14, 2014).

[25] Veith, G.D., DeFoe, D.L., Bergstedt, B.V., et al. 1979. Measuring and estimating the bioconcentration factor of chemicals in fish. *J. Fish Res. Board Can.* 36:1040–1048. As cited in: ATSDR, 2014. Agency for Toxic Substances and Disease Registry (ATSDR) 2013. *Toxicological Profile for Hexachlorobenzene. Draft for Public Comment - June.* Available at: http://www.atsdr.cdc.gov/toxprofiles/tp.asp?id=627&tid=115 (accessed January 14, 2014).

[26] United Nations Environmental Programme. 2008. *Stockholm Convention – Overview.* Available at: http://chm.pops.int/TheConvention/Overview/tabid/3351/Default.aspx (accessed January 18 2014).

[27] US Department of the Navy, Office of Naval Research. Undated. *Science and Technology Focus: Ocean in Motion: Currents – Characteristics.* Available at: http://www.onr.navy.mil/focus/ocean/motion/currents1.htm (accessed February 23, 2014).

[28] Rekacewicz, R. 1996. *World Ocean Thermohaline Circulation. UNEP/GRID-Arendal.* Available at: http://www.grida.no/graphicslib/detail/world-ocean-thermohaline-circulation_57ea (accessed February 9, 2014).

[29] US National Oceanographic and Atmospheric Administration. 2007. *NOAA's National Ocean Service: Education Currents: Waves.* Available at: http://oceanservice.noaa.gov/education/tutorial_currents/lessons/currents_tutorial.pdf (accessed February 9, 2014).

[30] Broecker, W.S. 1991. The great ocean conveyor. *Oceanography.* 4(2): 79–89.

[31] Ross, D. 1995. *Introduction to Oceanography.* New York: HarperCollins College, pp. 199–226, 339–343. As cited by National Oceanographic and Atmospheric Administration, 2007. *NOAA's National Ocean Service: Education Currents: Waves.* Available at: http://oceanservice.noaa.gov/education/tutorial_currents/lessons/currents_tutorial.pdf (accessed February 9, 2014).

[32] Ockenden, W.A., Breivik, K., Meijer, S.N., Steinnes, E., Sweetman, A.J., and Jones, K.C. 2003. The global re-cycling of persistent organic pollutants is strongly retarded by soils. *Environ.Pollution.* 121: 75–80.

[33] Zarfl, C., Scheringer, M., and Matthies, M. 2011. Screening criteria for long-range transport potential of organic substances in water. *Environ. Sci. Technol.,* 45: 10075–10081.

[34] United Nations Environmental Programme. 2006. *Report of the Persistent Organic Pollutants Review Committee on the Work of Its Second Meeting.* Addendum: Risk profile on perfluorooctane sulfonate Available at: http://chm.pops.int/TheConvention/ThePOPs/TheNewPOPs/tabid/2511/Default.aspx (accessed February 16, 2013).

[35] Sinclair, E. and Kannan, K. 2006. Mass loading and fate of perfluoroalkyl surfactants in wastewater treatment plants. *Environ. Sci. Technol.,* 40(5): 1408–1414.

[36] D'eon, J.C., Crozier, P.W., Furdui, V.I., Reiner, E.J., Libelo, E.L., and Mabury, S.A. 2009. Perfluorinated phosphonic acids in Canadian surface waters and wastewater treatment plant effluent: Discovery of a new class of perfluorinated acids. *Environ. Toxicol.Chem.* 28(10): 2101–2107.

[37] Arvaniti, O.S., Ventouri, E.I., Stasinakis, A.S., and Thomaidis, N.S. 2012. Occurrence of different classes of perfluorinated compounds in Greek wastewater treatment plants and determination of their solid–water distribution coefficients. *J. Hazard. Mater.* 239–240: 24–31.

[38] Campo, J., Masiá, A., Picó, Y., Farré, M., and Barceló, D. 2014. Distribution and fate of perfluoroalkyl substances in Mediterranean Spanish sewage treatment plants. *Sci. Total Environ.* 472: 912–922.

[39] Ahrens L., Felizeter, S., Sturm, R., Xie, Z., and Ebinghaus, R. 2009. Polyfluorinated compounds in waste water treatment plant effluents and surface waters along the River Elbe, Germany. *Marine Pollution Bull.* 58: 1326–1333.

[40] Yeung, L.W.Y, Nobuyoshi, Y., Sachi, T., Lam, P.K.S., Sinha, R.K., Borole, D.V., and Kannan, K. 2009. A survey of perfluorinated compounds in surface water and biota including dolphins from the Ganges River and in other waterbodies in India. *Chemosphere.* 76(1): 55–62.

[41] Guo, R., Sim, W.J., Lee, E.S., Lee, J.H., and Oh, J.E. 2010. Evaluation of the fate of perfluoroalkyl compounds in wastewater treatment plants. *Water Res.* 44(11): 3476–3486.

[42] Shivakoti, B.R., Tanaka, S., Fujii, S., Kunacheva, C., Boontanon, S.K., Musirat, C., Seneviratne, S.T.M.L.D., and Tanaka, H. 2010. Occurrences and behavior of perfluorinated compounds (PFCs) in several wastewater treatment plants (WWTPs) in Japan and Thailand. *J. Environ. Monitor.* 12(6):1255–1264.

[43] Kim, J.W., Tue, N.M., Isobe, T., Misaki, K., Takahasi, S., Viet, P.H., and Tanabe, S. 2013. Contamination by perfluorinated compounds in water near waste recycling and disposal sites in Vietnam. *Environ. Monitor. Assess.* 185(4): 2909–2919.

[44] Pan, Y., Shi, Y., Wang, J., and Cai, Y. 2011. Evaluation of perfluorinated compounds in seven wastewater treatment plants in Beijing urban areas. *Sci. China Chem.* 54(3): 552–558.

[45] Thompson, J., Eaglesham, G., Reungoat, J., Poussade, Y., Bartkow, M., Lawrence, M., and Mueller, J.F. 2011. Removal of PFOS, PFOA and other perfluoroalkyl acids at water reclamation plants in South East Queensland Australia. *Chemosphere.* 82(1): 9–17.

[46] Ahrens L., Felizeter, S., Sturm, R., Xie, Z., and Ebinghaus, R. 2009. Polyfluorinated compounds in waste water treatment plant effluents and surface waters along the River Elbe, Germany. *Marine Pollution Bull.* 58: 1326–1333.

[47] Ahrens, L., Gerwinski, W., Theobald, N., and Ebinghaus, A. 2010. Sources of polyfluoroalkyl compounds in the North Sea, Baltic Sea and Norwegian Sea: Evidence from their spatial distribution in surface water. *Marine Pollution Bull.* 60: 255–260.

[48] Theobald, N., Caliebe, C., Gerwinski, W., Huhnerfuss, H., and Lepom, P. 2011. Occurrence of perfluorinated organic acids in the North and Baltic seas. Part 1: Distribution in sea water. *Environ. Sci. Pollution Res.* 18(7): 1057–1069.

[49] Ahrens, L., Barber, J.L., Xie, Z., and Ebinghaus, A. 2009. Longitudinal and latitudinal distribution of perfluoroalkyl compounds in the surface water of the Atlantic Ocean. *Environ. Sci. Technol.* 43: 3122–3127.

[50] Ahrens, L., Xie, Z., and Ebinghaus, R. 2010. Distribution of perfluoroalkyl compounds in seawater from Northern Europe, Atlantic Ocean, and Southern Ocean. *Chemosphere.* 78:1011–1016.

[51] Cai, M., Zhao, Z., Yin, Z., Ahrens, L., Huang, P., Cai, M., Yang, H., et al. 2012. Occurrence of perfluoroalkyl compounds in surface waters from the North Pacific to the Arctic Ocean. *Environ. Sci. Technol.* 46(2): 661–668.

[52] Cai, M, Zhao, Z., Yang, H., Yin, Z., Hong, Q., Sturm, R., Ebinghaus, R., et al. 2012. Spatial distribution of per- and polyfluoroalkyl compounds in coastal waters from the East to South China Sea. *Environ. Pollution.* 161: 162–169.

[53] Yamashita, N., Taniyasu, S., Petrick, G., Wei, S., Gamo, T., Lam, P.K.S., and Kannan, K. 2008. Perfluorinated acids as novel chemical tracers of global circulation of ocean waters. *Chemosphere.* 70: 1247–1255.

[54] Yamashita, N., Kannan, K., Taniyasu, S., Horii, Y., Petrick, G., and Gamo, T. 2005. A global survey of perfluorinated acids in oceans. *Marine Pollution Bull.* 51(8): 658–668.

[55] OECD. 2002. *Co-operation on Existing Chemicals—Hazard Assessment of Perfluorooctane Sulfonate and its Salts.* ENV/JM/RD(2002)17/FINAL. Robust Study Report Reference No. 8 - Early Life-Stage Toxicity Test with the Fathead Minnow (*Pimephales promelas*), pp. 114-117. Available at: http://www.oecd.org/env/ehs/risk-assessment/2382880.pdf (accessed March 7, 2014).

[56] MacDonald, M.M., Warne, A.L., Stock, N.L., Mabury, S.A., Soloman, K.R., and Sibley, P.K. 2004. Toxicity of perfluorooctane sulfonic acid and perfluorooctanoic acid to *Chironomus tentans*. *Environ. Toxicol. Chem.* 23(9): 2116–2123. As cited in: United Nations Environmental Programme, 2006. *Report of the Persistent Organic Pollutants Review Committee on the Work of its Second Meeting.* Addendum: Risk profile on perfluorooctane sulfonate.

[57] Boudreau, T.M., Sibley, P.K., Mabury, S.A., Muir, D.C.G., and Solomon, K.R. 2003. Laboratory evaluation of the toxicity of perfluorooctane sulfonate (PFOS) on *Selenastrum capricornutum, Chlorella vulgaris, Lemna gibba, Daphnia magna* and *Daphnia pulicaria. Arch. Environ. Contam. Toxicol.* 44: 307–313. As cited in: United Nations Environmental Programme, 2006. *Report of the Persistent Organic Pollutants Review Committee on the Work of Its Second Meeting.* Addendum: Risk profile on perfluorooctane sulfonate.

[58] National Weather Service. 2011. *JetStream - Online School for Weather. The Jet Stream.* Available at: http://www.srh.noaa.gov/jetstream/global/jet.htm (accessed February 9, 2014).

[59] National Weather Service, 2011. *JetStream - Online School for Weather. Global Circulations and Jet Stream.* Available at: http://www.srh.noaa.gov/jetstream/global/circ.htm (accessed February 9, 2014).

[60] Committee on the Significance of International Transport of Air Pollutants; National Research Council. 2009. *Global Sources of Local Pollution: An Assessment of Long-Range Transport of Key Air Pollutants to and from the United States.* Washington, DC: National Academies Press, pp. 1–7; 18–23; 113–121. Appendix B. Available at: http://www.nap.edu/catalog.php?record_id=12743 (accessed February 2014).

[61] National Geographic Education. 2014. *Encyclopedic Entry: Jet Stream.* Available at: http://education.nationalgeographic.com/education/encyclopedia/jet-stream/?ar_a=1 (accessed February 9, 2014).

[62] Gouin, T., Mackay, D., Jones, K.C., Harner, T., and Meijer, S.N. 2004. Evidence for the "grasshopper" effect and fractionation during long-range atmospheric transport of organic contaminants. *Environ. Pollution.* 128: 139–148.

[63] von Waldow, H., MacLeod, M., Jones, K., Scheringer, M., and Hungerbühler, K. 2010. Remoteness from emission sources explains the fractionation pattern of polychlorinated biphenyls in the Northern Hemisphere. *Environ. Sci. Technol.* 44(16): 6183–6188.

[64] Mackay, D., Arnot, J.A., Webster, E., and Reid, L. 2009. The evolution and future of environmental fugacity models. In *Ecotoxicology Modeling, Emerging Topics in Ecotoxicology: Principles, Approaches and Perspectives 2.* Reid J. Devillers (Ed.), New York: Springer, pp. 355–375.

[65] US EPA. 2012. EPI Suite™ v4.11. *EPISuite User Guide.* Available at: http://www.epa.gov/oppt/exposure/pubs/episuitedl.htm (accessed January 4, 2014).

[66] OECD. 2004. *OECD SERIES ON TESTING AND ASSESSMENT No. 45. GUIDANCE DOCUMENT ON THE USE OF MULTIMEDIA MODELS FOR ESTIMATING OVERALL ENVIRONMENTAL PERSISTANCE AND LONG-RANGE TRANSPORT.* Available at: http://search.oecd.org/officialdocuments/displaydocumentpdf/?doclanguage=en&cote=env/jm/mono(2004)5 (accessed January 4, 2014).

[67] US EPA. 2000. *1,4-Dichlorobenzene (para-Dichlorobenzene).* Available at: http://www.epa.gov/ttn/atw/hlthef/dich-ben.html (accessed December 30, 2013).

[68] ECHA. 2013. *Committee for Risk Assessment (RAC)/Committee for Socio-economic Analysis (SEAC) - Background Document to the Opinion on the Annex XV dossier proposing restrictions on 1,4-DCB*. ECHA/RAC/RES-O-0000003486-69-01/F. ECHA/SEAC/RES-O-0000003486-69-02/F. Available at: http://echa.europa.eu/documents/10162/7b884aff-2852-40c3-8e7a-642ca92ca425 (accessed December 8, 2013).

[69] US EPA. 1980. *Sources of Toxic Compounds in Household Wastewater*. EPA-600/2-80-128. pp. 18, 39.

[70] Canadian Centre for Environmental Modelling and Chemistry. 2006. *Sewage Treatment Plant (STP) Model*. Version 2.11. Available at: http://www.trentu.ca/academic/aminss/envmodel/models/STP211.html (accessed December 31, 2013).

[71] US EPA 1982. *Fate of Priority Pollutants in Publicly Owned Treatment Works. Final Report*. Volume 1. EPA 440/1-82/303.

[72] US EPA. 2012. *EPI Suite™ v4.11*. Available at: http://www.epa.gov/oppt/exposure/pubs/episuitedl.htm (accessed December 31, 2013).

[73] European Communities. 2004. *1,4-DCB CAS No: 106-46-7, EINECS No: 203-400-5 RISK ASSESSMENT*. pp. 12. Available at: http://echa.europa.eu/documents/10162/fb7bf6b4-7831-4c3b-87b3-5acd493ce597 (accessed August 21, 2014).

[74] OECD. 2003. Descriptions of selected key generic terms used in chemical hazard/risk assessment. *OECD Series on Testing and Assessment*, Number 44. ENV/JM/MONO(2003)15. Available at: http://search.oecd.org/officialdocuments/publicdisplaydocumentpdf/?cote=ENV/JM/MONO(2003)15&docLanguage=En (accessed September 14, 2013).

[75] National Academy of Sciences. 1983. *Risk Assessment in the Federal Government: Managing the Process*. Committee on the Institutional Means for Assessment of Risks to Public Health, National Research Council. Available at: http://www.nap.edu/catalog/366.html (accessed September 14, 2013).

[76] European Chemicals Agency. 2009. *Guidance in a Nutshell - Chemical Safety Assessment*. Available at: http://echa.europa.eu/documents/10162/13632/nutshell_guidance_csa_en.pdf (accessed September 14, 2013).

[77] Ministry of Environmental Protection (MEP) P.R. China. 2012. *The Guidelines for Risk Assessment of Chemicals, China (Draft)*. Translated by REACH24H Consulting Group. eBook07 2012/01/20.

[78] Slovik, P. 1999. Trust, emotion, sex, politics, and science: Surveying the risk-assessment battlefield. *Risk Anal.* 19(4): 689–701.

[79] OECD. 2000. *Manual for Investigation of HPV Chemicals. Chapter 3: Data Evaluation*. Available at: http://www.oecd.org/env/ehs/risk-assessment/49191971.pdf (accessed September 14, 2013).

[80] ECHA. 2009. *Practical Guide 6: How to Report Read-Across and Categories*. Available at: http://echa.europa.eu/documents/10162/13655/pg_report_readacross_en.pdf (accessed September 14, 2013).

[81] US EPA. 2010. *TSCA New Chemicals Program (NCP) Chemical Categories*. Available at: http://www.epa.gov/oppt/newchems/pubs/npcchemicalcategories.pdf (accessed September 14, 2013).

[82] OECD. 2012. Assessment of chemicals. Chapter 2. *Data Gathering and Testing: SIDS, the SIDS Plan and the SIDS Dossier*. Last updated 2012. Available at: http://www.oecd.org/chemicalsafety/risk-assessment/chapter2datagatheringandtestingsidsthesidsplanandthesidsdossier.htm (accessed October 20, 2013).

[83] ECHA. 2011. *Guidance on Information Requirements and Chemical Safety Assessment.* Part C: PBT Assessment. Version 1.1. Available at: http://echa.europa.eu/documents/10162/13643/information_requirements_part_c_en.pdf (accessed October 20, 2013).

[84] ECHA. 2012. Guidance on information requirements and chemical safety assessment. Chapter R.8: *Characterization of Dose [Concentration]-Response for Human Health.* Version 2.1. Available at: http://echa.europa.eu/documents/10162/13632/information_requirements_r8_en.pdf (accessed October 20, 2013.

[85] US EPA. 2012. *Integrated Risk Information System (IRIS): Basic Information.* Web page last updated September 26, 2012. Available at: http://www.epa.gov/iris/intro.htm (accessed October 20, 2013).

[86] US EPA. 1998. Guidelines for ecological risk assessment. Published on May 14, 1998, *Federal Register.* 63(93): 26846–26924. Available at: http://www.epa.gov/raf/publications/pdfs/ECOTXTBX.PDF (accessed October 20, 2013).

[87] Jackson, M., Lemus-Olalde, R., Inhof, C., Yin, Z., Locey, B., Magee, B.H., and Connor, K.T. 2013. Differences between US EPA iris inhalation reference concentrations (RfCs) and European Chemicals Agency (ECHA) long-term inhalation derived no effect levels (DNELs) for the general population. *Toxicologist.* 132(1): 418.

[88] OECD. 1999. *OECD Series on Testing and Assessment* Number 17. Environmental Exposure Assessment Strategies for Existing Industrial Chemicals in OECD Member Countries. pp. 11–14. Available at: http://search.oecd.org/officialdocuments/displaydocumentpdf/?doclanguage=en&cote=env/jm/mono(99)10 (accessed January 4, 2014).

[89] ECHA. 2010. *Guidance on Information Requirements and Chemical Safety Assessment Part B: Hazard Assessment.* Draft new chapter B.8 Scope of Exposure Assessment. Table 1. Available at: http://echa.europa.eu/documents/10162/13632/inforeq_csr_b_draft_update_peg_2_web_en.pdf (accessed January 4, 2014).

[90] OECD. 2013. *OECD Guidelines for Testing of Chemicals - Full List of Test Guidelines.* July 2013. Available at: http://www.oecd.org/env/ehs/testing/TG_List_EN_Jul_2013.pdf (accessed January 18, 2014).

[91] Hunt, R.G., Sellers, J.D., and Franklin, W.E. 1992. Resource and environmental profile analysis: A life cycle environmental assessment for products and procedures. *Environ. Impact Assess. Rev.* 12(3): 245–269.

[92] Hunt, R.G. and Franklin, W.E. 1996. LCA- How it came about - personal reflections on the origin and LCA in the USA. *Int. J. Life Cycle Assess.* 1(1): 4–7.

[93] United Nations Environmental Programme. 2011. *Towards a Life Cycle Sustainability Assessment: Making Informed Choices on Products.* Available at: http://www.unep.org/publications/contents/pub_details_search.asp?ID=6236 (accessed November 1, 2013).

[94] International Standards Organisation. 2006. ISO 14044: *2006 - Environmental Management— Life Cycle Assessment— Requirements and Guidelines.*

[95] European Commission Joint Research Centre. Undated. *Our Thinking – Life Cycle Thinking.* Available at: http://lct.jrc.ec.europa.eu/index_jrc (accessed January 12, 2014).

[96] US EPA. 2013. *Life Cycle Perspective.* Available at: http://www.epa.gov/nrmrl/std/lifecycle.html (accessed January 12, 2014).

[97] US EPA. 2006. *Life Cycle Assessment: Principles and Practice*. Prepared by Scientific Applications International Corporation for US Environmental Protection Agency, Washington, DC, EPA/600/R-06/060. Available at: http://cfpub.epa.gov/si/si_public_record_Report.cfm?dirEntryID=155087 (accessed November 1, 2013).

[98] Heijungs, R. and Guinee, J.B. 2012. An overview of the life cycle assessment method – Past, present, and future. In *Life Cycle Assessment Handbook: A Guide for Environmentally Sustainable Products*, Mary Ann Curran (Ed.), Beverly, MA: Scrivener, pp. 15–30.

[99] US EPA. 2014. *LCA Resources – Software and Databases*. Web page last updated January 21, 2014. Available at: http://www.epa.gov/nrmrl/std/lca/resources.html#Software (accessed January 26, 2014).

[100] Ciroth, A. 2011. Software for life cycle assessment. In *Life Cycle Assessment Handbook: A Guide for Environmentally Sustainable Products*, Mary Ann Curran (Ed.), Beverly, MA: Scrivener, pp. 15–30.

[101] Chemical Abstracts Service. 2008. *CAS® Statistical Summary 1907–2007*. Columbus, OH: Chemical Abstracts Service.

[102] International Council of Chemical Associations. 2012. *ICCA Responsible Care® Progress Report - Growing Our Future*. Available at: http://www.icca-chem.org/ICCADocs/RC%20annual%20report.pdf (accessed December 26, 2013).

[103] Carson, R. 1962. *Silent Spring*. Boston: Mariner Books, Houghton Mifflin Harcourt.

[104] President's Council on Environmental Quality. 1971. Council on Environmental Quality, Toxic Substances, reprinted in *Staff House Committee on Interstate and Foreign Commerce*, 94th Congress, 2nd Session, Legislative History of the Toxic Substances Control Act (TSCA Legislative History) at 760 (Comm. Print 1976), pp. 759–760.

3

Product Stewardship Regulations

Reports of chemicals in the environment or in our bodies can provoke strong emotional responses. Some believe that the widespread distribution of low levels of chemicals is an acceptable price to pay for the convenience and safety afforded by the use of many chemicals; others may blame chemical companies for an apparent focus on profits at the expense of the environment. But the explanation is more nuanced than that and depends in part upon the regulations that societies choose to impose on the manufacture, import, and use of chemicals.

This chapter examines the regulatory environment in three major economies, the United States (US), European Union (EU), and China.* Together these three geographies account for most of the world's chemical production. Figure 3.1 illustrates the value of chemical production in those geographies, beginning at a time coincident with the first chemical control regulations and extending through the present [3,4]. The United Nations Environment Programme (UNEP) reports figures that illustrate the growth of the global chemical industry during this period: worth approximately $171 billion USD† in 1970, the value grew to $4.12 trillion USD by 2010. UNEP further cites studies that suggest that the chemical market will grow globally by approximately 3% per year until 2050, with the growth in China anticipated to be especially dramatic [3].

A comprehensive regulatory review would examine restrictions on chemical substances, pesticides, pharmaceuticals, cosmetics, food additives, and food contact materials. Such an examination would fill several volumes. This book narrows the focus to regulations on chemical substances used to manufacture various industrial and consumer products. The history of those programs provides much of the explanation for current environmental conditions.

* National environmental laws often reflect two global treaties, the Rotterdam Convention and the Stockholm Convention. Brief summaries follow. The Rotterdam Convention establishes a prior informed consent (PIC) procedure for importing shipments of chemicals listed in Annex III of the convention and provides for information exchange between participating countries. Adopted in September 1998, it entered into force February 2004 [1]. The 2001 Stockholm Convention requires parties to the treaty to take measures to eliminate or reduce the release of persistent organic pollutants (POPs) into the environment [2]. As described in this chapter, regulations on persistent, bioaccumulative, and toxic substances address POPs.

† Industry value in 1970 dollars; adjusted for inflation using the US Department of Labor Bureau of Labor Statistics inflation calculator at http://www.bls.gov/data/inflation_calculator.htm, equivalent to $1.01 trillion in 2010.

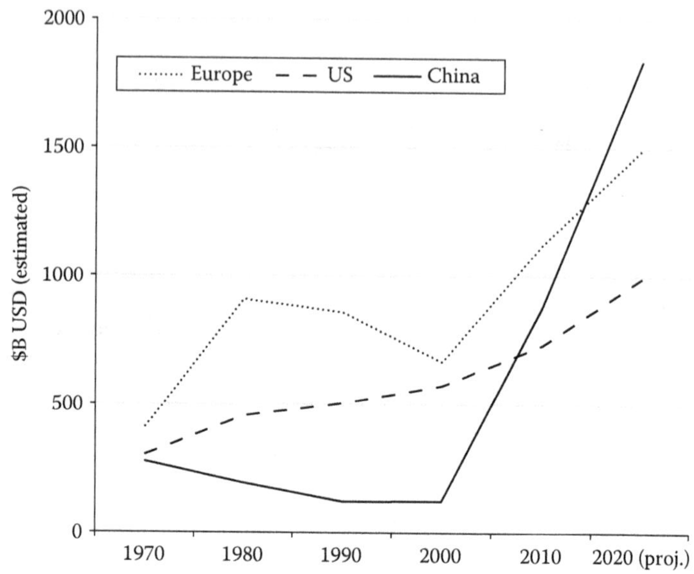

FIGURE 3.1
Chemical production in the United States, Europe, and China. *Note: Estimates for 1970–2000 adjusted to 2010 dollars.*

Most modern chemical control regulations reflect the fundamental premise that some chemical releases to the environment are inevitable, and that the use of chemicals should be controlled so that those releases present an acceptable level of risk. That simple premise leads to challenging technical and ethical questions. What is "acceptable" risk? How does one calculate that risk? How does a society choose to balance the possible risks against the apparent benefits of using a chemical? How is the potential risk from a single new chemical put into the context of exposure to multiple chemicals?

This chapter explores those questions, beginning with a discussion of one of the oldest chemical control laws, the US Toxic Substances Control Act. Readers should note that the discussions of laws and regulations that follow are intended simply to illuminate the question of why chemicals can be found throughout our environment and should not be construed as legal or regulatory advice.

3.1 United States

"We should no longer be limited to repairing damage after it has been done; nor should we continue to allow the entire population or the entire environment to be used as a laboratory" [5, p. 760]. With this straightforward declaration of purpose in its 1971 report *Toxic Substances*, the President's Council on Environmental Quality introduced the need for chemical control legislation in the United States. The story of how the Toxic Substances Control Act (TSCA) grew from such origins illustrates the forces that influenced the development of product stewardship legislation.

3.1.1 How Did We Get Here?

Early environmental laws in the United States controlled gross environmental pollution in specific media. Initial efforts focused on surface water, beginning with the Rivers and Harbors Act of 1899 and culminating in the Clean Water Act of 1972. The Clean Air Act of 1972 limited air pollution. The Safe Drinking Water Act of 1974 allowed the federal government to regulate the quality of public water supplies. The primary product-focused legislation passed by the midpoint of the twentieth century, the Federal Insecticide, Fungicide, and Rodenticide Act (FIFRA), required manufacturers to register pesticide products with the US Department of Agriculture (USDA) before interstate or international shipment and to label products with adequate warnings and precautionary instructions for use. But no regulation governed the manufacture or importation of chemical substances in general.

In fact, when President Richard Nixon's Advisory Council on Executive Organization recommended a structure for a new "Environmental Protection Administration" in 1970, it focused almost entirely on the organizational structure and function needed for pollution control and pesticide regulation [6]. The mission of the soon-to-be Environmental Protection Agency (EPA) had no part in evaluating or regulating most of the 62,000 chemicals then in commerce, either as raw materials or in products other than pesticides, drugs, or food additives.

Recognition of this gap was growing, however. In 1970, the President's Council on Environmental Quality began to study the need for chemical legislation. The Council reached three stark conclusions [5]:

- Toxic substances are entering the environment.
- These substances can have severe effects.
- Existing legal authorities are inadequate.

In consequence, the Council recommended new legislation in a report that expressed the objectives of that new law with remarkable passion [5, p. 762]:

> The approach called for in the new Toxic Substances Control Act is a new way of looking at environmental problems. Rather than dealing with pollutants as they appear in air, in water, and on land, it represents a systematic and comprehensive approach to the problem. It relies on understanding the flow of potentially toxic substances throughout the entire range of activity—from extraction to production to consumer use and to

> disposal. Only through such a comprehensive approach can we provide protection to man and his environment. In the last few years, we have identified the enormity of the problem; we have developed the institutional capability through the creation of EPA to look comprehensively at pollution of the environment. The time has come for an action program to control the use of toxic substances.

In essence, the Council proposed to anchor the regulation of chemicals in life cycle thinking.

Once the Council made its recommendations, the specific provisions of a bill had to be negotiated among the Congress, the newly formed US EPA, the Commerce Department, the chemical industry, and other stakeholders [7]. Those negotiations, as one would expect, reflected not just the emerging environmental health science but the beliefs and the economic conditions of the time. The record of the debate in Congress reflects concerns that the proposed TSCA would impose burdensome costs and would harm industry [8]. This debate occurred during a period when the American economy experienced tremendous strain. Figure 3.2 [9,10] shows the fluctuation in two key indicators of the national economy between 1970, when the President's Council on Environmental Quality began to study the need for chemical legislation, and 1976, when TSCA became law. Most notably, in October 1973 Arab members of the Organization of Petroleum Exporting Countries (OPEC) imposed an oil embargo against the United States to protest American support of Israel [11]. OPEC cut oil production and banned petroleum exports to the United States and other countries. Oil prices skyrocketed and the US economy faltered. This strain on the economy permeated national thinking.

3.1.2 1976 Toxic Substances Control Act

The Act passed in 1976 represented a compromise between environmental protection and industry concerns. The White House expressed this balance in a press release issued when President Gerald Ford signed the bill [12]:

> Generally speaking, the bill gives authority to the EPA Administrator to:
>
> - Require private industry to provide test data and supply detailed information on specified substances;
> - Prevent, or place limitations on, the marketing of new substances which the Administrator believes harmful; and,
> - Ban or limit continued marketing of existing substances....
>
> The Toxic Substances Control Act is designed to prevent problems. By allowing early and selective regulation of only those uses that are likely to be hazardous, the Act minimized adverse regulatory impacts on the chemical industry. In addition, this preventative approach should help reduce the need for regulations under other laws which hurt important industries such as fishing, food processing, and the many other manufacturers who rely on chemical products.

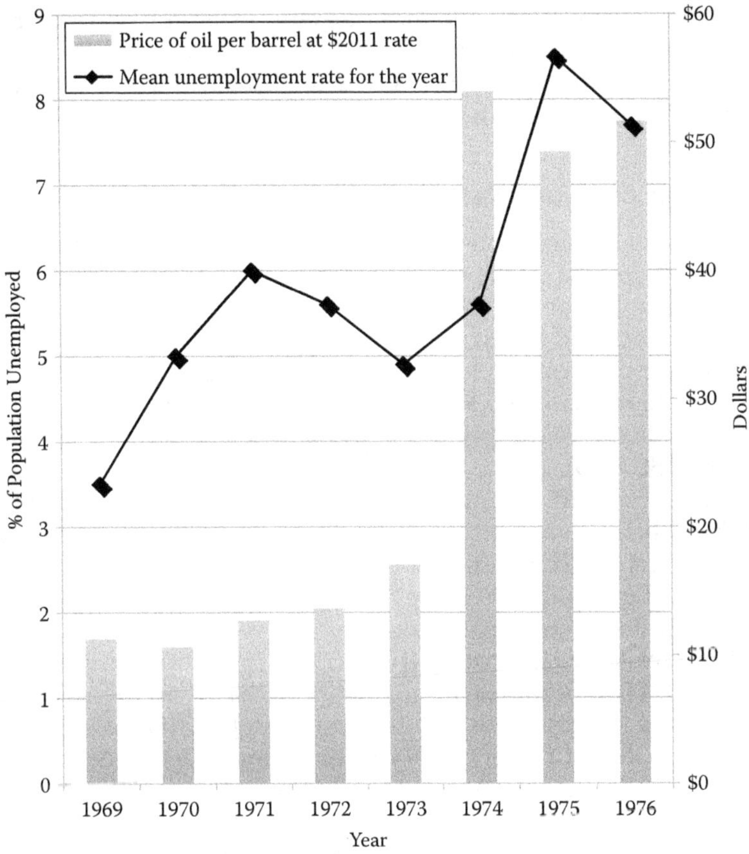

FIGURE 3.2
Indicators of US economy during TSCA debate.

The crucial distinction between new and existing chemical substances* and the focus on "only those uses that are likely to be hazardous" expressed in this press release set the course for the regulation of chemicals in the United States for the next 40 years.

The President's Council on Environmental Quality catalyzed the development of TSCA with the finding that toxic substances were entering the environment and that those substances could have severe effects. In hindsight, was TSCA adequate to address those concerns? To address that question, the provisions of TSCA pertaining to existing chemicals and new chemicals are each discussed below. These narratives reflect the evolution of the program

* A chemical substance is, with some exemptions, "... any organic or inorganic substance of a particular molecular identity, including—(i) any combination of such substances occurring in whole or in part as a result of a chemical reaction or occurring in nature, and (ii) any element or uncombined radical" [13].

over the past 40-some years, and include examples or milestones from specific time periods to illustrate key points.

Let's start by looking at how the agency implemented the TSCA program for the chemical substances that existed when regulation began.

3.1.2.1 Existing Chemical Substances

Regulators faced a tremendous challenge when the law was passed in 1976. First, the newly formed EPA needed to build the infrastructure for a new regulatory program [18]. Second, TSCA required immediate action on multiple fronts. The agency had to inventory existing chemicals at the same time as it was receiving notifications for some 700 to 1,000 new substances per year and developing regulations for three substances (polychlorinated biphenyls (PCBs), chlorofluorocarbon propellants, and dioxin wastes) as specifically required by TSCA [19].

Chemical substances that were manufactured in or imported into the United States before January 1, 1975 were considered to be existing chemicals under TSCA [7]. The initial effort simply to inventory those chemicals, which included promulgation of the necessary regulations and collection of information, lasted three years. The original TSCA Inventory listed approximately 62,000 existing chemicals [14]. The TSCA Inventory has grown to list more than 84,000 existing chemical substances, based on the submittal of premanufacture notices of the manufacture (or importation) of new chemicals. (Protocols for notifying the US EPA of the manufacture or import of a new chemical substance are described in the next section of this chapter.)

It is important to recognize that not all of the substances listed on the TSCA Inventory are actually in commerce or used in large quantities. Under the Chemical Data Reporting (CDR) rule, the US EPA periodically collects information from industry on the manufacture and importation of chemicals. In 2012, companies reported that they manufactured or imported a total of 7,674 chemicals in amounts of 25,000 pounds or more at each of their sites during calendar year 2011, not including exempt substances. (In 2006, companies reported 6,200 chemicals [15].) Approximately 3,000 chemicals are manufactured or imported in quantities of 1,000,000 pounds per year [16,17].

The discussion of the regulation of existing substances that follows describes:

- Overview of requirements
- Characterization of existing chemicals
- Regulation of existing chemicals
- Accomplishments

3.1.2.1.1 Overview of Requirements

TSCA gives the US EPA the authority to collect and develop data on existing chemicals, and to regulate those chemicals that present an unreasonable risk. In brief, statutory authorities under TSCA include (but are not limited to) the following provisions:

- Section 8: The US EPA can request data on the properties, manufacture, exposure, and health and environmental effects of existing chemicals. Companies must keep records of adverse effects from their chemicals on health or the environment. Notably, under Section 8(e), responsible parties must report information that reasonably supports the conclusion that a substance or mixture presents a substantial risk of injury to health or the environment.
- Section 4: The US EPA may require manufacturers, importers, and processors to test chemicals when risks or exposures of concern are identified. This requires a formal rule making; alternatively, the US EPA has worked with industry to develop data through Enforceable Consent Agreements (ECAs) and Voluntary Testing Agreements (VTAs) [20].
- Section 6: If an "unreasonable risk of injury to health or the environment" from a chemical substance has been proven, the US EPA can require risk-abatement action such as labeling chemical substances, regulating uses, restrictions on disposal, and prohibiting or limiting manufacture. Section 6(e) requires the EPA to regulate PCBs, and Titles 2, 3, and 4 regulate the production, importation, use, and disposal of asbestos, radon, and lead-based paint. Additionally, Section 6 regulates fully halogenated chlorofluoroalkanes (CFCs) as described in the Code of Federal Regulations (CFR), Part 40, Section 762 (40 CFR 762).

Of these key provisions, certain aspects of Sections 8 and 6 are most relevant to the central questions of this book. Paraphrasing TSCA, in order to regulate one of the tens of thousands of existing chemicals, the US EPA must have information that shows that the substance presents an unreasonable risk. Let's look at the programs designed to collect information about the existing chemicals in greatest use, and then at efforts to regulate "unreasonable risk."

3.1.2.1.2 *Characterization of Existing Chemicals: High Production Volume Chemicals and Work Plan Chemicals*

The US EPA defines high production volume (HPV) chemicals as those chemicals produced or imported in the United States in quantities of 1,000,000 pounds or more per year. The HPV Challenge program began with the recognition in 1998, some 20 years after passage of TSCA, that no data were publicly available on the basic toxicity of 43% of HPV chemicals. Further, only 7% of HPV chemicals had data available for six fundamental toxicity testing endpoints [21]. Consequently, the US EPA developed a voluntary program in which manufacturers and importers of HPV chemicals sponsored the development of screening-level health and environmental data. The program included approximately 2,200 existing chemicals [22].

As of July 2013, the US EPA had amassed over 340 submissions representing almost 900 chemical substances, either as a single chemical submission or as a member of a chemical category, and made the information publicly available in an online database known as the High Production Volume Information System (HPVIS). The agency has evaluated the toxicity data for 229 of these substances/categories and published a report on each to prioritize HPV chemicals for follow-up data collection or management actions based upon their potential risks [23]. These reports "consist of an evaluation of the quality and completeness of the data set provided in the Challenge Program submissions. They are not intended to be definitive statements regarding the possibility of unreasonable risk of injury to health or the environment" [24, p. 1]. Thus, they do not provide direct insight into the potential that the US EPA may take regulatory action on any HPV chemical.

In summary, approximately 2,200 substances are HPV chemicals. The US EPA has collected data on approximately 900 substances and assessed the data for approximately one-quarter of those substances. The agency plans to further assess and may regulate some of these chemicals as described below.

In 2012, the US EPA announced a renewed initiative on existing chemicals [25]. Pending the potential development of new legislation, the agency intended to refocus on a three-pronged approach to managing existing chemicals:

1. Risk assessment and risk reduction
2. Data collection and screening
3. Public access to chemical data and information

The US EPA identified 83 Work Plan chemicals for further risk assessment in 2012 and revised the list in 2014 to comprise 90 substances [26]. A subset of those chemicals (or groups of chemicals), 15 as of 2014, has been selected for action from among:

- HPV chemicals
- Chemicals identified as persistent, bioaccumulative, and toxic (PBT)
- Chemicals in consumer products
- Chemicals potentially of concern for children's health because of reproductive or developmental effects
- Chemicals subject to review and potential action in international forums
- Chemicals found in human biomonitoring programs
- Chemicals in categories generally identified as being of potential concern in the new chemicals program

The US EPA has stated that it may take actions on certain chemicals as a consequence of this program. It may initiate regulatory action to label, restrict, or ban a chemical, or to require the submission of additional data needed to determine the risk [27]. Should the agency choose to do so, it will use its authority under Section 6 of TSCA.

3.1.2.1.3 *Regulation of Existing Chemicals*

Section 6 is the "hammer" provision of TSCA: if a chemical presents an unreasonable risk, then the US EPA can restrict or ban the substance. Using the term "unreasonable risk" to describe the impetus for action understates the complexity of determining that a chemical should be regulated. According to Section 6(c) of TSCA, the agency must consider:

> (A) the effects of such substance or mixture on health and the magnitude of the exposure of human beings to such substance or mixture,
> (B) the effects of such substance or mixture on the environment and the magnitude of the exposure of the environment to such substance or mixture,
> (C) the benefits of such substance or mixture for various uses and the availability of substitutes for such uses, and
> (D) the reasonably ascertainable economic consequences of the rule, after consideration of the effect on the national economy, small business, technological innovation, the environment, and public health.

The EPA must also consider whether "a risk of injury to health or the environment could be eliminated or reduced to a sufficient extent by actions taken under another Federal law (or laws)" administered by the agency.

If, after this evaluation, the US EPA concludes that a substance may present an "unreasonable risk," then under TSCA Section 6(a) (emphasis added),

> … the Administrator shall by rule apply one or more of the following requirements to such substance or mixture to the extent necessary to protect adequately against such risk <u>using the least burdensome requirements</u>:
>
> (1) A requirement
> (A) prohibiting the manufacturing, processing, or distribution in commerce of such substance or mixture, or
> (B) limiting the amount of such substance or mixture which may be manufactured, processed, or distributed in commerce.
> (2) A requirement
> (A) prohibiting the manufacture, processing, or distribution in commerce of such substance or mixture for
> (i) a particular use or
> (ii) a particular use in a concentration in excess of a level specified by the Administrator …, or

(B) limiting the amount of such substance or mixture which may be manufactured, processed, or distributed in commerce for [particular uses]...

(3) A requirement that such substance or mixture or any article containing such substance or mixture be marked with or accompanied by clear and adequate warnings and instructions with respect to its use, distribution in commerce, or disposal or with respect to any combination of such activities. .....

(4) A requirement that manufacturers and processors of such substance or mixture make and retain records of the processes used to manufacture or process such substance or mixture and monitor or conduct tests which are reasonable and necessary to assure compliance with the requirements of any rule applicable under this subsection.

To see how these criteria apply in practice, let's examine the history of efforts to regulate asbestos.

In the first three years of the TSCA program, the agency deliberately chose to give low priority to Section 6 regulatory actions. Officials reportedly believed that most existing substances for which data were available were (or could be) adequately controlled under other laws and that resources should therefore be focused on reviewing new chemicals and on developing data on existing chemicals [28]. But regulators were concerned enough about the possible effects of exposure to asbestos to make action under Section 6 a priority.

In 1979, the US EPA published an Advance Notice of Proposed Rulemaking (ANPRM) describing how it intended to use its new authority under TSCA to "prevent and reduce any unreasonable risks that are identified" from exposure to asbestos [29]. The ANPRM described the agency's plan to analyze health risks throughout the life cycle of asbestos; perform a socioeconomic and substitute assessment; and, if appropriate based on those analyses, regulate asbestos. The US Government Accountability Office framed the concern that prompted this ANPRM in remarkable terms: "Asbestos, a proven carcinogen, may be a chemical time-bomb for the thousands of persons that were (and are being) exposed to it" [19, p. 25].

After six years of analysis, the US EPA proposed a rule in January 1986 to regulate asbestos. Public comment and legislative hearings followed. In 1989, three years after publication of the proposed rule, the agency issued a final regulation that would "prohibit, at staged intervals, the future manufacture, importation, processing, and distribution in commerce of asbestos in almost all products" and require some product labeling, although the rule allowed for some exemptions. The *Federal Register* notice of this rulemaking reported that the regulation was based on "over 45,000 pages of analysis, comments, testimony, correspondence, and other materials ... [and] the extensive record developed by [the Occupational Safety and Health Administration] OSHA ..." [30].

The US Government Accountability Office described what happened next [31, p. 9]:

> EPA spent 10 years exploring the need for the asbestos ban and developing the regulation. On the basis of its review of over 100 studies of the health risks of asbestos as well as public comments on the proposed rule, EPA determined that asbestos is a potential carcinogen at all levels of exposure—that is, that it has no known safe exposure level. EPA's 1989 rule under TSCA Section 6 prohibited the future manufacture, importation, processing, and distribution of asbestos in almost all products. In response, some manufacturers of asbestos products filed suit against EPA arguing, in part, that the rule was not promulgated on the basis of substantial evidence regarding unreasonable risk. In October 1991, the U.S. Court of Appeals for the Fifth Circuit agreed with the manufacturers, concluding that EPA had failed to muster substantial evidence to justify its asbestos ban. Specifically, the court concluded that EPA did not consider all necessary evidence and failed to show that the control action it chose was the least burdensome regulation that would adequately protect human health or the environment. EPA had not calculated the risk levels for intermediate levels of regulation because it believed there was no asbestos exposure level for which the risk of injury or death was zero. As articulated by the court, the proper course of action for EPA, after an initial showing of product danger, would have been to consider each regulatory option, beginning with the least burdensome, and the costs and benefits of each option. The court further criticized EPA's ban of products for which no substitutes were currently available stating that, in such cases, EPA "bears a tough burden" to demonstrate, as TSCA requires, that a ban is the least burdensome alternative.

The results of this case have reportedly discouraged the US EPA from further efforts to utilize Section 6 [31–33].

3.1.2.1.4 Accomplishments: Existing Chemicals

As briefly described above, the US EPA has used the powers granted to it under TSCA to collect information about existing chemicals and to regulate them. The numbers tell the story. Over 40 years after the President's Council on Environmental Quality recommended passage of an act that would control the use of toxic substances, there are, according to the references cited above:

- 62,000 chemicals potentially in commerce that existed when Congress created TSCA
- 7,674 chemicals* manufactured or imported in amounts of 25,000 pounds or more per site
- ~2,200 HPV chemicals manufactured or imported at >1,000,000 pounds per year

* Includes both chemicals on the initial Inventory and those subsequently added. Does not include substances formally exempted from TSCA notification.

- 900 HPV substances characterized in HPVIS
- ~229 HPV substances that have undergone screening level hazard characterization
- 90 Work Plan chemicals or groups of chemicals
- 15 Action Plan chemicals or groups of chemicals
- 5 existing chemicals or categories of chemicals regulated under Section 6 due to "unreasonable risk"*

In short, only a fraction of the chemicals in commerce when Congress passed the law are regulated. The debate over whether additional "existing chemicals" should be regulated, which could easily fill several more books, reignited when Congress considered the Chemical Safety Improvement Act of 2013. (See Section 3.1.3 for more information on that proposed act.)

The provisions of TSCA for existing substances account for the regulation of over 62,000 of the chemicals on the TSCA Inventory. What, then, of the more than 24,000 new chemical substances put into commerce in the United States since TSCA passed in 1976?

3.1.2.2 New Chemical Substances

Section 5 of TSCA requires a manufacturer or importer of a new chemical substance to notify the US EPA before beginning manufacture or importation. The agency reviews the premanufacture notice (PMN) and may require additional testing or restrict the conditions under which a chemical is manufactured or used through a Consent Order and potentially a Significant New Use Rule (SNUR). This regulatory process is so crucial to the life cycle of chemicals in the United States that it is worth a closer look. The subsections that follow describe these aspects of TSCA requirements for new chemicals:

- Premanufacture notice and exemptions
- US EPA's review of premanufacture notices, tools to estimate exposure and hazard, and criteria for risk assessment during that review
- Regulation of new chemicals
- Outcomes

3.1.2.2.1 Premanufacture Notice and Exemptions

The notification form in use as of 2014 [35] begins with information about the substance identity and its manufacture and use. The PMN contains two fundamental types of information relevant to this discussion: exposures (i.e., human exposure and environmental release) and data on the physical/

* These five chemicals or categories of chemicals are: polychlorinated biphenyls (PCBs), fully halogenated chlorofluoroalkanes, dioxin, asbestos, and hexavalent chromium. (Note that a portion of the US EPA's 1991 asbestos rule stands [34].)

chemical and toxicological properties of the substance. The submitter can claim information in the notice as confidential.

To characterize the potential for human exposure and environmental release, the PMN submitter must describe occupational exposure, including the estimated number of workers exposed to the substance, the activities during which they would be exposed, and any protective equipment or engineering controls that would reduce exposure. Estimates of environmental release and exposure must include estimates of releases to air, surface water, land, incineration, or other media. The PMN form also requests optional pollution prevention information. The submitter may provide information on how the proposed manufacturing process or use of the new chemical substance contributes to source reduction, recycling activities, and safer products or processes. The US EPA takes this information into account when assessing potential risks.

To allow the US EPA to consider potential risks from exposure, the PMN must also include "all existing health and environmental data in the possession of the submitter, parent company, or affiliates, and a description of any existing data known to or reasonably ascertainable by the submitter" (40 CFR 720.40(d)). TSCA does not specify a minimum "base set" of required data and PMNs have often contained few data. In a survey of the PMNs submitted between 1979 and 1985, the US EPA found that 44% of the 5,500 submittals contained some toxicological data (typically on acute toxicity); 9% contained some ecotoxicological data (again, typically on acute toxicity), and only 6% contained information on biodegradability. Fifty-four percent contained no test data. Trends circa 1997 were similar [36].

Herein lies the central paradox of the 1976 law: The US EPA must demonstrate that a new chemical substance poses an "unreasonable risk" before it can request specific test data on the substance [37,38]. The point bears repeating: A manufacturer or importer does not need to test the toxicity or ecotoxicity of a new chemical substance under the 1976 TSCA, and the US EPA cannot demand testing to obtain those data unless it can demonstrate—without necessarily having any supporting data on which to base that demonstration—that a substance could present a substantial risk due to its toxicity and the likely extent of exposure. After a brief discussion of the exemptions to the notification requirement, we examine how the agency performs that evaluation.

Manufacture of a new chemical substance does not always require a PMN. A manufacturer or importer may be exempt from the requirement under certain circumstances. TSCA regulations specify five exemptions, most of which pertain in general terms to circumstances where exposure to the substance would be limited:

- *Low Volume Exemption* (LVE) may apply if 10,000 kilograms or less of the substance will be manufactured or imported each year under the requirements at 40 CFR 723.50.

- A substance manufactured in small quantities for research and development may be eligible for the *Research and Development* exemption if special procedural and recordkeeping requirements are met, as specified at 40 CFR 720.36 and 720.78.
- If the manufacture and use of a substance is so well controlled that exposure will be minimal, the *Low Releases and Low Exposures* (LoREX) exemption may apply under 40 CFR 723.50.
- If a new substance is manufactured or imported for test marketing, the *Test Marketing Exemption* (TME) specified at 40 CFR 720.38 may apply.
- The *Polymer Exemption* may apply under conditions defined at 40 CFR 723.250.

3.1.2.2.2 Agency Review of PMN

The US EPA reviews the PMN to determine whether a new chemical substance, in the words of Section 4 of TSCA, "may present an unreasonable risk of injury to health or the environment." If so, the agency can impose controls on the manufacture or use of the substance; it can also require testing if significant exposure to a chemical could occur. The PMN assessment considers the potential for exposure throughout the substance's life cycle and the risk to human health or environmental organisms from such exposure. The agency must complete this review within 90 days, and consequently a carefully controlled and scheduled review process has been developed [36].

The assessment may be streamlined if the new substance falls into one of approximately 60 chemical categories. The US EPA developed these categories to make the review process more efficient, with the recognition that some new chemical substances could efficiently be evaluated by analogy with structurally similar chemicals with shared chemical and toxicological properties. Before the agency developed the category approach in 1987, approximately 20% of PMNs underwent detailed review; based on the category approach, the US EPA reviews only 2 to 3% of PMNs in detail [39].

This task brings us back to the TSCA paradox: Monitoring data on exposures to the chemical during its manufacture and use may be lacking for a new chemical manufactured only at pilot scale, and as described above, a PMN need not contain a minimum set of toxicology data. Consequently, the agency uses various tools to estimate physicochemical properties, approximate exposures, and assess the potential toxicity of a substance in the absence of data. Without data, each estimation step introduces uncertainty into the evaluation. The description of the tools used in these steps is intended to give the reader a sense of how regulators gauge the potential for environmental release and effects from a new chemical in order to determine whether it must be regulated, and the potential for uncertainties in those estimates. It is not intended to provide a detailed working knowledge of each of the models.

Physical/chemical data reported in the PMN may provide a basis for the mass balance calculations that allow one to estimate the concentrations

TABLE 3.1

EpiSuite™ Modules

Module	Parameter Estimated
MPBPWIN™	Melting point, boiling point, and vapor pressure
WATERNT™	Water solubility
WSKOWWIN™	Water solubility
KOWWIN™:	Log octanol–water partition coefficient, log K_{ow}
HENRYWIN™	Henry's law constant
KOAWIN	Octanol–air partition coefficient
LEV3EPI™	Level III multimedia fugacity model predicts partitioning of chemicals among air, soil, sediment, and water under steady-state conditions for a default model "environment"
AOPWIN™	Gas-phase reaction rate and chemical atmospheric half-lives
AEROWIN™	Fraction of airborne substance sorbed to airborne particulates, that is, the parameter phi (φ)
WVOLWIN™	Rate of volatilization of a chemical from rivers and lakes
HYDROWIN™	Aqueous hydrolysis rate constants and half-lives for certain substances
BIOWIN™	Aerobic and anaerobic biodegradability
BioHCwin	Biodegradation half-life for hydrocarbons
STPWIN™	Removal of a chemical in a typical activated sludge-based sewage treatment plant (total removal and three processes that may contribute to removal: biodegradation, sorption to sludge, and air stripping)
BCFBAF™	Fish bioconcentration factor
KOCWIN™	Organic carbon-normalized sorption coefficient for soil and sediment, that is, KOC

released to air, water, and land. Absent those physical/chemical data, the agency can use models based on structure–activity relationships (SAR) to estimate critical parameters. One such model is the Estimation Programs Interface (EPI) Suite™, discussed briefly in Chapter 2. The US EPA's Office of Pollution Prevention and Toxics and Syracuse Research Corporation (SRC) developed this screening-level tool, which contains a series of modules that estimate various properties and environmental fate and transport processes based on the chemical's structure. Table 3.1 lists the modules in version 4.11 of EPISuite™ and their output [40] to illustrate the range and types of data that may be estimated.

These estimated parameters can be used to model the behavior of a new chemical substance. The inaccuracies inherent in deriving physical/chemical properties or fate and transport behavior from SAR, rather than from reliable experimental data, will obviously affect the accuracy of exposure predictions.

A PMN contains the submitter's estimates of exposure. To verify such estimates or to develop exposure scenarios a PMN lacks, the US EPA may turn to various mass balance techniques or models [41,42]. The simplest exposure assessment technique relies on generic exposure scenarios. The *1997 Chemical Engineering Branch Compilation of Generic Scenarios (for Industry-Specific Workplace*

Release and Exposure Estimation) [43] comprises approximately 60 dossiers on chemical manufacturing processes that the US EPA prepared between 1991 and 1995. These dossiers illustrate a relatively early basis, 15 to 20 years into the TSCA program, for estimating the release of chemicals during manufacturing.

The 1991 generic scenario for the manufacture of soap and detergent provides one example that illustrates how relatively little quantitative information was available in the early years of the TSCA program [44]. That scenario presents assumptions related to exposures to a PMN substance, apparently used as a fragrance, for example, in household cleaners. The US EPA would use these assumptions about exposures to evaluate the proposed risks from the manufacture and use of a new chemical unless the PMN provided more specific information. The assumptions included estimates of the concentration of the substance and the amount used in total per site; the number of workers exposed; the amount of the substance that would be released to the air; and the daily dermal exposure. Further, "[w]ater releases should be limited to equipment cleanout." In other words, the generic scenario does not discuss the potential for a soap fragrance to be discharged into wastewater during routine use of the product, and then to a publicly owned treatment works (POTW) and potentially to surface water. Of the 11 quantitative assumptions presented in the generic scenario, no reference or basis was provided for 6 assumptions and partial references (e.g., author and year but not title of publication) were provided for 5 assumptions. This example illustrates some of the challenges in accurately characterizing exposures during the manufacture and use of a new chemical, particularly during the earlier days of the TSCA program.

In addition to relying on the generic exposure scenarios, the US EPA has developed models to estimate exposures. The models ChemSTEER™ and E-FAST™ [45] can provide screening-level mass balance estimates. These models are briefly described below. Higher-tier tools are also available to evaluate some specialized applications [45].

ChemSTEER [45,48], which incorporates some of the generic exposure scenarios discussed above, estimates occupational inhalation and dermal exposure to a chemical during industrial and commercial manufacturing, processing, and use operations, and can also estimate the associated releases to air, water, and land. The software performs a mass balance based on information about the chemical, its physical/chemical properties, and the processes it undergoes. The US EPA can use this screening-level tool during the review of a PMN to estimate workplace releases and exposures conservatively, and has noted in the model documentation [46, p. 3] that "of the new chemicals regulated under Section 5, a significant fraction are regulated based on estimated workplace release and exposure."

Early environmental exposure assessments focused on releases to surface water under the premise that releases to terrestrial ecosystems or through landfills to the environment were negligible [48]. The US EPA utilized the percentile stream flow method and the probabilistic dilution model (PDM) to estimate the concentrations of a new chemical substance in the environment. The analysis began with engineers at the agency who estimated the amounts of chemical releases (kilograms per day per site, and estimated days per year), as well as the degree of wastewater treatment such releases would likely undergo. Modelers then estimated the concentration of that release, known as the predicted environmental concentration (PEC), as it was instantaneously mixed with the flow in a stream. In a screening-level analysis, the percentile stream flow method predicted multiple PECs for various stream flows to represent a range of hypothetical real-world conditions, for example, 50th percentile mean stream flow, 50th percentile low stream flow, 10th percentile mean stream flow, and 10th percentile low stream flow. The PDM method offered more refined predictions, including the number of days per year that a target concentration might be exceeded.

As the TSCA program continued to evolve, modeling efforts became more sophisticated. The PDM was incorporated into software, the Exposure and Fate Assessment Screening Tool or E-FAST, that predicts multimedia exposures. The current E-FAST model provides conservative, screening-level estimates of the concentrations of chemicals released to air, surface water, and landfills, and from consumer products. The US EPA developed the program to support the risk assessment of both new and existing chemicals. Using input information regarding the physical/chemical properties and chemical fate of a substance, the model generates estimates of inhalation and dermal and ingestion dose rates for human receptors via various exposure routes. It can also estimate the number of days per year that a concentration of concern to an aquatic receptor would be exceeded [45,49].

In addition to these programs, the EPISuite software contains two modules used in exposure assessment [40]. LEV3EPI™ is a Level III multimedia fugacity model that predicts partitioning of organic chemicals among air, soil, sediment, and water under steady-state conditions for a default model "environment." STPWIN™ models the removal of an organic chemical in a typical activated sludge-based sewage treatment plant, calculating total removal and the fraction by each of three processes that may contribute to removal: biodegradation, sorption to sludge, and air stripping.

Table 3.2 indicates the criteria by which the US EPA determines that exposure, whether estimated from these models or based on information provided in the PMN, may be significant. If the manufacture or use of a new chemical substance could meet or exceed these criteria, then the agency may require exposure-based testing of toxicity or ecotoxicity in a consent order negotiated under TSCA §5 [50,51,52].

The US EPA has generally defined the types of data needed to assess the risks from potentially significant exposures as shown in Table 3.3 [53]. The agency has also defined the testing that will typically be required for new chemical substances within certain categories [54].

TABLE 3.2

Criteria for Significant Exposure Under US EPA's Exposure-Based Policy

Exposure Parameter	TSCA 5(e) Exposure-Based Policy Criterion
Production volume	100,000 kg/yr
Significant or substantial human exposure: high number of workers exposed	≥1,000 workers
Significant or substantial human exposure: acute worker exposure, inhalation	≥100 workers exposed to ≥10 mg/day
Significant or substantial human exposure: chronic worker exposure, inhalation	≥100 workers exposed to 1–10 mg/day for ≥100 days/yr
Significant or substantial human exposure: chronic worker exposure, dermal	≥250 workers exposed by routine dermal contact for ≥100 days/yr
Significant or substantial human exposure: consumer	Presence in consumer product where exposures are likely
Significant human exposure: ambient general population	≥0.003 mg/kg/day exposure via drinking water, air, or groundwater
Substantial human exposure: ambient general population	≥10,000 kg/yr release to environmental media
Substantial environmental release	≥1,000 kg/yr total release to surface water calculated after wastewater treatment

As noted above, submitters often provide few data with a PMN. In such cases [50, p. 27],

> The paucity of human, animal, and aquatic toxicity data for most PMN substances has led [U.S. EPA] scientists to use several different approaches for hazard identification. These approaches include: consideration of the likelihood of absorption from the lung, gastrointestinal tract, and skin; consideration of the expected products of metabolism and their toxicity; structure-activity relationships (SARs); and consideration of the presence of structural groups or substituents that are known to bestow toxicity.

Scientists may evaluate SARs using their knowledge of the chemical and biochemical behavior of various molecular structures. They may also use various models to explore SARs, some of which are described below.

Given information on the structure of a new chemical substance, the Analog Identification Methodology (AIM) Engine™ software analyzes the structure based on over 700 molecular identifiers and matches the substance with potential analogues from a database of over 86,000 substances. One can then access publicly available data on the analogous substances to understand the potential hazards of exposure to the new chemical substance [55].

The PBT Profiler software determines, as the name implies, if a chemical could be considered to be persistent, bioaccumulative, or toxic. The profiler highlights chemicals at two apparent levels of persistence: those with a half-life in water, soil, or sediment between two and six months, and those with

TABLE 3.3

Testing Triggered Under Exposure-Based Policy by Potentially Significant Exposure

Exposure Concern	Possible Testing Requirements
Significant human exposure	Bacterial reverse mutation (formerly "Ames") test (OPPTS 870.5100/OECD 471) Mammalian erythrocyte micronucleus (intraperitoneal route) (OPPTS 870.5395/OECD 474) Repeated dose 28-day oral toxicity in rodents (OPPTS 870.3050 or OECD 407) Acute oral toxicity (OPPTS 870.1100/OECD 401, or OECD 425 ["Acute Oral Toxicity: Up-and-Down Procedure"])
New substance produced at >1,000,000 pounds (approximately 454,000 kilograms), and for which high exposures to workers or exposure to any consumers or the general population is expected	In addition to the tests listed above, "Prenatal Developmental Toxicity" test (OPPTS 870.3700; oral route, one species), OECD 421 ("Reproduction/Developmental Toxicity Screening Test") or OECD 422 (combined OECD 407 (28-day study) and OECD 421).
Substantial environmental release	Algal toxicity (OPPTS 850.5400/OECD 201 ("Alga, Growth Inhibition Test")) • Aquatic invertebrate acute toxicity, freshwater daphnids (OPPTS 850.1010/OECD 202 ("*Daphnia* sp. • Acute Immobilisation Test and Reproduction Test")) • Fish acute toxicity (OPPTS 850.1075/OECD 203) Potentially, chronic toxicity testing
PMN substances that meet the drinking water, ground water, total release to environment media, or total release to surface water criteria	• Water solubility—OPPTS 830.7840 or 835.7860; OECD 105 • Fate in wastewater treatment—Porous Pot test (OPPTS 835.3220); or Semi-Continuous Activated Sludge test (OPPTS 835.3210; OECD 302A); or Modified Zahn-Wellens/EMPA test (OPPTS 835.3200; OECD 302B); or Activated Sludge Sorption Isotherm test (OPPTS 835.1110) • Aerobic biodegradation—Ready Biodegradability test (OPPTS 835.3110; OECD 301A-F); or Sealed Vessel CO2 Production test (OPPTS 835.3120) • Anaerobic biodegradation—Anaerobic Biodegradability of Organic Chemicals (OPPTS 835.3400) • Soil biodegradation—OPPTS 835.3300; OECD 304A • Photolysis (photodegradation)— OPPTS 835.2210 • Hydrolysis as a function of pH (stability in water)— OPPTS 835.2110 or 835.2130, or OECD 111

an apparent half-life greater than six months. A substance is considered to be bioaccumulative if the bioconcentration factor (BCF) is estimated to fall between 1,000 and 5,000, and very bioaccumulative if the BCF exceeds 5,000. Finally, the profiler indicates if the chemical belongs to a chemical category developed under TSCA. The determination that a new chemical substance may be PBT can trigger the need for additional testing and, potentially, control [56,57].

The US EPA's OncoLogic™ software estimates the potential for a chemical to cause cancer in humans. It incorporates SAR analysis and a functional analysis that reflects knowledge of how chemicals cause cancer in animals and humans [58].

The US EPA uses the *NonCancer Screening Protocol* as one method for screening chemicals—including a new chemical substance described in a PMN—for noncancer health effects in the absence of toxicity data [59]. The protocol comprises the following steps:

1. Locate measured data on chemical/analogue, including physical/chemical properties and chemical and biochemical reactions.
2. Determine if chemical/analogue has familiar and well-understood structure(s), by comparison to the chemical categories discussed previously and to lists of chemicals causing local effects or chemicals causing systemic effects within the protocol.
3. Search online for measured data.
4. Use screening models, appropriately applied, to predict data.
5. Have a toxicologist review data and estimate concern level, that is,
    - High concern (evidence of adverse effects in humans or conclusive evidence of severe effects in animal studies).
    - Moderate concern (suggestive animal studies, analogue data, chemical class known to produce toxicity).
    - Low concern (no concern identified).

The ECOSAR module of EPISuite, which was described above, estimates the aquatic toxicity of organic substances using quantitative structure activity relationships (QSARs). It can predict the acute toxicity to fish, daphnids, and green algae, and certain other endpoints for some chemical classes [60].

With actual data or estimates of the exposures and toxicity in hand, the US EPA may proceed to assess the potential risks. The US GAO framed that process circa 2001 as follows [61, p. 91]:

> Although TSCA gives EPA general authority to seek out and regulate any "unreasonable risk" associated with new or existing chemicals, there are two major limitations on the agency's regulatory actions. First, as implemented by EPA, regulation under TSCA involves consideration

> of both risks and applying the least burdensome requirement needed to regulate the risk. The term "unreasonable risk" is not defined in TSCA.
>
> However, according to EPA, the legislative history indicates that unreasonable risk involves the balancing of the probability that harm will occur, and the magnitude and severity of that harm, against the effect of a proposed regulatory action on the availability to society of the expected benefits of the chemical substance. The second major limitation on EPA's authority under TSCA is a requirement to defer to other federal laws. Generally, if a risk of injury to health or the environment could be eliminated or reduced to a sufficient extent by actions taken under another federal law, that other law must be deferred to unless it can be shown to be in the public interest to regulate under TSCA …
>
> [N]ewly introduced chemicals (or uses) can be regulated under TSCA based on whether they may present an unreasonable risk, and this finding of risk can be based on data for structurally similar chemicals. Because industrial chemicals in commerce in 1975–1977 were "grandfathered" into the inventory without considering whether they were hazardous, there are situations in which existing chemicals might not be controlled, while EPA would act to control a new chemical of similar or less toxicity under the PMN program.

The agency's challenge is immense: Within a 90-day period, the US EPA must review a PMN (that need not contain data) to determine whether exposure to that substance might cause an "unreasonable risk," the magnitude of which is not defined by law and which must be balanced against other considerations. The US EPA has not published explicit guidance on how it reaches a finding of "unreasonable risk" but has described it generally as follows [62]:

> EPA's determination that manufacture, processing, use, distribution in commerce, or disposal of an individual substance which has been the subject of a notice under section 5 of the TSCA may present an unreasonable risk of injury to human health or the environment is based on consideration of (i) the size of the risks identified by EPA; (ii) limitations on risk that would result from specific safeguards (generally, exposure and release controls) sought based on Agency review; and (iii) the benefits to industry and the public expected to be provided by new chemical substances intended to be manufactured after Agency review. In considering risk, EPA considers factors including environmental effects, distribution, and fate of the chemical substance in the environment, disposal methods, waste water treatment, use of protective equipment and engineering controls, use patterns, and market potential of the chemical substance.

This approach contrasts sharply with agency actions under a program like Superfund, where the criteria for significant risk at uncontrolled hazardous waste sites have been clearly quantified. Under that program, for example, regulations at 40 CFR 300.430(e)(2)(i)(A) specify that remediation goals shall consider the following:

> For systemic toxicants, acceptable exposure levels shall represent concentration levels to which the human population, including sensitive subgroups, may be exposed without adverse effect during a lifetime or part of a lifetime, incorporating an adequate margin of safety;
>
> For known or suspected carcinogens, acceptable exposure levels are generally concentration levels that represent an excess upper bound lifetime cancer risk to an individual of between 10^{-4} and 10^{-6} using information on the relationship between dose and response. The 10^{-6} risk level shall be used as the point of departure for determining remediation goals. …

Lacking explicit regulatory provisions or a transparent decision-making process under TSCA, one can still infer something about the US EPA's assessment of the risk from exposure to a new chemical substance by triangulating from the three published benchmarks described below.

Below certain levels of exposure, the TSCA program presumes that risk is likely to be limited. Table 3.2 illustrates that general philosophy, as does the LoREX exemption, which provides the first of the three benchmarks for acceptable exposures. Federal regulations specify the LoREX criteria at 40 CFR 723.50(2). Paraphrased, exposure to releases of a new chemical substance during its manufacture and use are generally considered to be negligible when there would be:

- No dermal exposure or inhalation exposure (except for limited releases from incinerators) by consumers or the general population
- No exposure in drinking water greater than a 1 milligram per year (mg/yr) estimated average dosage
- No dermal or inhalation exposure by workers, allowing for the use of exposure controls
- No releases resulting in surface water concentrations above 1 part per billion (ppb), unless a higher concentration can be supported
- No releases of the new substance from incineration at a level above 1 microgram per cubic meter (mg/m^3) maximum average concentration
- No releases to land or groundwater or to a landfill unless the substance has negligible groundwater migration potential

This exemption cannot be invoked if the substance, any reasonably anticipated metabolites, environmental transformation products, or byproducts of the substance, or any reasonably anticipated impurities in the substance may cause, under anticipated conditions of manufacture, processing, distribution in commerce, use, or disposal of the new chemical substance, serious acute or chronic human health effects or significant environmental effects.

Although this exemption does not define serious human health effects or significant environmental effects to a degree that helps to explain the concept of "unacceptable risk" under TSCA, it does provide a point of reference: The risks from exposure to any compound at 1 mg/yr in drinking water, 1 ppb in surface water, or 1 mg/m^3 in air are anticipated to be acceptable. These levels are quite low, but compare the following information relevant to the drinking water criterion of 1 mg/yr. A risk assessor working under US EPA guidelines typically assumes that an adult drinks 2 liters of water per day [63]. Over the course of a year, an adult could therefore drink water containing up to 0.0014 milligrams per liter (mg/L) of a chemical substance before reaching the LoREX criterion of 1 mg/year of exposure. Of the 76 (nonradionuclide) chemicals the US EPA regulates under the Safe Drinking Water Act, the agency has promulgated maximum contaminant levels for 10 substances that are below 0.0014 mg/L [64]. In other words, the US EPA's risk assessment of 13% of the chemicals regulated under the National Primary Drinking Water Regulations found that exposure to 1 mg/yr in drinking water (which is considered to be a negligible exposure under the TSCA New Chemicals program) was not acceptable.

A second set of benchmarks for understanding what the US EPA may consider to be an unreasonable risk pertains to the potential risk to human health from exposure to a new chemical. Here we have two reference points: the first, a publication by an agency official early in the TSCA program regarding the determination of acceptable risks under TSCA, and the second, the US EPA's explanation of how it derives limits for worker exposure to new chemicals under TSCA.

In a 1983 discussion of the US EPA's risk management strategy under TSCA, a US EPA official indicated [65, p. 164] that, with respect to noncancer health risks, the US EPA's point of departure for risk-management decisions was essentially zero risk. The no observed adverse effect level (NOAEL) based on animal data was adjusted using an uncertainty factor to determine the allowable daily intake (ADI). With respect to carcinogenic risks, the objective was to reduce risks (generally by reducing exposure) to an "insignificant" level. However, the official noted that the US EPA did "not employ any predetermined statistical risk level since this will vary depending on a variety of factors." This lack of specificity differs markedly from the contemporaneous focus in the Superfund program on a target risk reduction to an excess upper bound lifetime cancer risk to an individual of between 10^{-4} and 10^{-6}.

New chemical exposure limits (NCELs) for new chemicals regulated under TSCA are modeled after OSHA's permissible exposure limits (PELs) for workers exposed to a substance by inhalation. NCELs may include an eight-hour time weighted average limit or a short-term exposure limit for a chemical substance. In 1995, the US EPA described the risk basis for determining

NCELs under TSCA as follows [66], noting that different protocols are used to derive limits based on noncarcinogenic and carcinogenic effects.

Tests for toxic effects other than cancer typically determine a NOAEL or lowest observed adverse effect level (LOAEL). To set an NCEL for a noncarcinogenic effect, the US EPA divides the NOAEL or LOAEL by an uncertainty factor. That factor reflects the type of study and its duration; for example, a factor of 100 is typically used to adjust the data from a developmental toxicity study. The quotient is considered a permissible daily dose, that is, the dose at which an adult human would be unlikely to suffer adverse effects if exposed for a working lifetime. The agency derives the NCEL from the permissible daily dose by accounting for the typical body weight of an adult, the volume of air inhaled during a typical eight-hour work shift, and in some cases, by an absorbtion factor.

To derive an NCEL for a potential carcinogen, the US EPA begins with the policy that a cancer risk of 10^{-4} (one case of cancer in a population of 10,000 exposed workers) is acceptable. The corresponding NCEL is calculated from the cancer slope factor, which describes the dose–response for the chemical under consideration, incorporating assumptions about the number of workdays in a year and number of years worked in a lifetime, as well as typical body weight of an adult, and the volume of air inhaled during a typical eight-hour work shift.

In summary, the NCEL for a new chemical substance may be derived from a level anticipated to cause no adverse noncarcinogenic effects or to present a risk of perhaps one cancer in every 10,000 workers exposed to the substance over their working lifetimes. A new chemical can be released into the workplace atmosphere at concentrations up to the NCEL without any engineering or exposure controls required. The US EPA acknowledges that in some cases the NCEL concentrations initially calculated from these methods may be difficult to attain or monitor, and that in some cases feasibility must be taken into consideration when setting an NCEL. One can infer that in such cases the potential risk would be higher than the benchmarks described above. Those hoping to deconstruct an NCEL to understand the underlying data and inherent risk management decisions may be stymied: Under TSCA, the toxicity data used in the agency's risk deliberations on a new chemical substance may be submitted as confidential business information [35] and therefore not publicly available.

The third benchmark relates to potential effects on the ecosystem. The US EPA gauges the potential effects of a release of a new chemical substance on the aquatic ecosystem by calculating a concern concentration (CC) and comparing it to the PEC. The CC is that concentration, which, if exceeded, may cause a significant risk to aquatic organisms. The CC for each chemical is determined by applying assessment factors (AsF) or uncertainty factors (UF) to the effect concentrations determined from aquatic toxicity testing or QSAR calculations. These factors account for uncertainties such as the test species' sensitivity or laboratory conditions, for example. If the PEC is below the CC, the probability of a significant risk occurring is low [67–69].

3.1.2.2.3 *Regulation of New Chemicals*

After assessing the potential risks from a new chemical, the US EPA may decide to regulate its importation, manufacture, or use. Under Section 5(e) of TSCA, the agency can issue a consent order to the manufacturer or importer or promulgate a SNUR. Section 5(f) gives the agency an alternate form of authority to manage a new chemical substance. Each of these regulatory mechanisms is discussed briefly below.

If the agency determines that under certain conditions a new substance may present an unreasonable risk, it can issue a consent order to the PMN submitter that requires one or more of the following [70,71]:

- Testing for environmental fate and/or toxicity parameters
- Worker protection measures, including
  - Use of personal protective equipment
  - NCELs
  - Hazard communication language
- Restrictions on
  - Distribution and use
  - Releases to water, air, and land
- Recordkeeping

When the US EPA believes the manufacture or use of a new chemical should be restricted, the agency promulgates a SNUR under Section 5. The SNUR may bind a manufacturer or importer to certain handling precautions, such as the use of personal protective equipment or prohibition of discharge to surface water, for example. Any "significant new use" requires a significant new use notification (SNUN) that allows the US EPA the opportunity to review the use and, if necessary, to take action to limit exposures.

Finally, the US EPA can regulate a new chemical under Section 5(f) of TSCA, which says in part that if:

> ... there may be a reasonable basis to conclude that a chemical substance or mixture presents or will present a significant risk of serious or widespread harm to human beings from cancer, gene mutations, or birth defects, the Administrator shall, within the 180-day period beginning on the date of the receipt of such data or information, initiate appropriate action under section 5, 6, or 7 to prevent or reduce to a sufficient extent such risk or publish in the Federal Register a finding that such risk is not unreasonable.

TABLE 3.4
New Chemical Substances, 1979–September 30, 2010

Action	Number
PMNs submitted	36,623
PMNs withdrawn	–1,848
Exemptions	13,826
Total new chemicals	*48,601*
Section 5(e) Consent Order, with or without SNUR	1,492
Non-5(e) SNURs	797
Section 5(f) regulation	4
Total new chemicals regulated[a]	*2,293*

[a] Does not reflect restrictions implicit in certain exemptions such as the LoREX exemption (81 substances, 1979–2013).

The agency has utilized this provision once to date. In 1984 the US EPA proposed a rule, which still stands, to regulate four new substances in metalworking fluids: mixed mono and diamides of an organic acid, triethanolamine salts of a substituted organic acid, triethanolanime salt of tricarboxylic acid, and tricarboxylic acid. The rule prohibits mixing these substances with nitrosating agents, as the US EPA determined that the resulting reaction could produce a carcinogen that would pose a significant risk. (See 40 CFR 747 – Metalworking Fluids; also [72].)

3.1.2.2.4 Outcomes: New Chemical Substances

As of 2013, the US EPA has screened on the order of 40,000 PMNs and received over 10,000 exemption requests. The disposition of the PMNs submitted in the first 30 years of the program, shown in Table 3.4, illustrates how those PMNs are often handled [73,74].

3.1.2.3 Implications and Calls for Reauthorization

Readers who have persisted through this discussion of the byzantine provisions of TSCA may well be confused at this point in the narrative. Regulators working under TSCA to govern the introduction of new chemical substances into the marketplace and the use of existing chemicals have worked within a law that restricts the agency's ability to collect new data, regulates chemicals based on "unreasonable risk" without defining the term, and requires that risks be balanced against the effect of a proposed regulatory action on the availability to society of the expected benefits of the chemical substance. Consequently, the manufacture and use of chemical substances in the United States since 1976 reflects a series of risk-management decisions often, of necessity, based on limited data. TSCA has regulated only a small percentage of the chemicals in commerce in the United States. Some argue that is sufficient or nearly so. Others echo the conclusions of the President's Council on Environmental Quality in 1970, believing that despite regulation under TSCA, toxic substances are entering the environment, these substances can have severe effects, and existing legal authorities are inadequate. In consequence, debate about reauthorizing TSCA intensified in 2013 with the introduction of the Chemical Safety Improvement Act.

3.1.3 TSCA Reform

Efforts to reform TSCA kindled in 2010 with the introduction of the Toxic Chemicals Safety Act of 2010 into the House of Representatives and the Safe Chemicals Act of 2010 into the Senate. Those efforts failed to gain momentum, however.

In 2013, the Chemical Safety Improvement Act (CSIA, S. 1009) was introduced into the Senate [178]. This proposed bill differed markedly from TSCA in several respects. Although debate over the legislation continues as of this writing, the provisions of the CSIA may indicate the direction that chemical control law will take in the United States. These provisions include the following:

- Revisions to the process for evaluating chemical risks and the standard by which such risks would be evaluated. CSIA would require the US EPA to make a safety determination on a chemical, or, to paraphrase the proposed bill, to ensure that no unreasonable risk of harm to human health or the environment would result from exposure to a chemical under the intended conditions of use.
- The CSIA would require the US EPA to develop a "framework" for evaluating chemicals. The bill would require the agency to establish a risk-based screening process for identifying existing chemical substances that are considered a "high priority" for a safety assessment and determination and a "low priority" for a safety assessment and determination.
- New chemical substances would be notified under a modified protocol for PMN submittal and review.
- The US EPA would have increased authority to require chemical testing, for example, based on the need to make a safety determination, rather than limiting the authority (as it is currently under TSCA) to require testing to cases where a chemical may present an unreasonable risk or has substantial potential for exposure.
- If a chemical did not meet the safety standard, then the use of the chemical could be restricted via labeling and warning requirements, recordkeeping requirements, data development obligations,

production limits, and bans or phaseouts of a chemical use or distribution method. Economic and benefit factors would be considered in assessing potential restrictions.

The proposed CSIA contains other hotly debated provisions regarding preemption of state regulations, the management of confidential business information, and deadlines for certain actions. Debate over the bill continued as of late 2014 [179].

In 2014, while the Senate was discussing the CSIA, the Chemicals in Commerce Act (CICA) was introduced into the House of Representatives. But debate stalled and the bill was not expected to pass in 2014 [180].

3.1.4 Recapitulation

- TSCA, passed in 1976, distinguished the 62,000 existing chemicals from chemicals newly placed into commerce after the law came into effect.
- TSCA did not require manufacturers or importers to provide data on existing chemicals. Section 8 of TSCA enables the US EPA to request data on the properties, manufacture, exposure, and health and environmental effects of existing chemicals. Section 8 also requires responsible parties to report information that reasonably supports the conclusion that a substance or mixture presents a substantial risk of injury to health or the environment. Under Section 4 of TSCA, the US EPA may require manufacturers, importers, and processors to test chemicals if the agency determines that risks or exposures may be of concern.
- The agency has largely relied on voluntary submittal of data on existing chemicals, notably though the HPV program. The US EPA has collected data on approximately 900 substances and assessed the data for approximately one quarter of them. In addition, the US EPA has identified 90 Work Plan chemicals for further assessment.
- Section 5 of TSCA requires companies to notify the US EPA when they begin to manufacture or import a new chemical, unless an exemption applies. The notification must include all available data but the law does not mandate testing. If the US EPA determines from the available data or by analogy to another substance that a new chemical could pose a risk, the agency can request test data or impose certain requirements on manufacture or import.
- The agency has developed numerous models to assess exposure and toxicity in the absence of data in order to determine whether a new chemical should be regulated. The criteria for determining when a risk would be unreasonable are not specified in the law, regulation, or written policy. As a consequence of the confidentiality of much of the data that are submitted and the unwritten policy on

risk findings, regulatory decisions have not been particularly transparent. Based on submittals between 1979 and September 30, 2010, approximately 6% of new chemicals were regulated.

- Section 6 of TSCA allows the US EPA to require risk-abatement action if an "unreasonable risk of injury to health or the environment" from a chemical substance has been proven. Five chemicals or categories of chemicals have been regulated under Section 6.
- Legislative efforts to reform TSCA circa 2010–2014 reexamined key provisions including the US EPA's authority to require test data, and the standard and methodology for assessing risk. As of this writing, in late 2014, the debates continued.

3.2 European Union and Member States

Chemical regulations in Europe developed as the European Union was evolving. The formation of the European Union is far outside the scope of this book, but a brief overview of recent history sets the stage for understanding the development of product stewardship regulations. The outline below documents milestones in the coalescence of the European Union after World War II [75] to provide context for the development of regulations on the manufacture and use of chemicals.

- 1952: European Coal and Steel Community (ECSC) formed by France, Germany, Italy, and the Benelux countries in order to support the free movement of coal and steel and free access to sources of production.
- 1957: Germany, France, Italy, the Netherlands, Belgium, and Luxembourg sign the Treaties of Rome. The first Treaty created the European Economic Community (EEC), or common market, to allow for the free movement of people and trade across national borders. The second created the European Atomic Energy Community (Euratom).
- 1965: The Treaty of Brussels, known as the Merger Treaty, merged the EEC, the ECSC, and Euratom under the Council of Ministers and European Commission. The merged entity was known as the European Communities (EC).
- 1970s: Four Member States – Denmark, Ireland, the United Kingdom, and Greece – joined the European Communities.
- 1980s: Spain and Portugal joined the European Communities.
- 1992: The European Union was formed with the signature of the Maastricht Treaty, which formalized cooperation in foreign policy,

defense, police and justice, and monetary union within the existing Communities (Euratom, ECSC, EEC). The EEC became the European Community (EC).

- 1994: Austria, Finland, and Sweden joined the European Communities.
- 2000s: The number of Member States increased to 28 (as of 2014) with the addition of Croatia, Cyprus, Estonia, Hungary, Latvia, Lithuania, Malta, Poland, the Czech Republic, Slovakia, Slovenia, Bulgaria, and Romania.

3.2.1 Origins of Chemical Regulations

Regulation of dangerous substances began in the European Communities in 1967, with the adoption of Council Directive of 27 June 1967 in the Approximation of Laws, Regulations, and Administrative Provisions Relating to the Classification, Packaging and Labelling of Dangerous Substances (67/548/EEC). This Directive defined *dangerous substances* as those which were explosive, oxidizing, easily flammable, flammable, toxic, harmful, corrosive, or irritant. But the initial provisions did not require testing or notification of chemical substances, simply that they be appropriately packaged and labeled. This limited scope has been attributed to the limitations of the Treaty of Rome which then bound together the EEC: Directive 67/548/EEC was based on articles in the Treaty intended to protect the health and safety of European workers and required each Member State to adopt uniform Community-wide laws rather than laws that imposed barriers to trade within the EEC. Chemicals could only be regulated within those limitations [76].

The manufacture and use of chemicals were not regulated for another decade, and then only in narrowly defined instances. Council Directive of 27 July 1976 on the approximation of the laws, regulations and administrative provisions of the Member States relating restrictions on the marketing and use of certain dangerous substances and preparations (76/769/EEC) allowed the authorities to restrict substances for particular uses. The original Directive restricted only PCBs and chloro-1-ethylene, but grew to regulate approximately 900 substances after a series of amendments.

The second major regulatory initiative was the Sixth Amendment to the Dangerous Substances Directive. Spurred by concerns over the risks from PCBs and by the development of domestic legislation in France, discussion of expanding the Dangerous Substances Directive to include controls on the marketing and use of dangerous chemicals began in 1974. That discussion accelerated after the United States passed TSCA in 1976, in consideration of the impact of that legislation on EEC exports to the United States [76]. Those discussions ultimately resulted in the passage of the Sixth Amendment to the Dangerous Substances Directive, which required testing of new chemical substances and information exchange between member nation-states.

The subsequent Existing Substances Regulation (EEC 793/93) specified the requirements for evaluating risks using such data and, if necessary, reducing risks.

The critical provisions of these regulations related to the control of chemical manufacture and use are discussed below.

3.2.2 Limitations Directive

Council Directive of 27 July 1976 on the approximation of the laws, regulations and administrative provisions of the Member States relating to restrictions on the marketing and use of certain dangerous substances and preparations (76/79/EEC), also known as the Limitations Directive, addressed two concerns [77]. First, authorities believed that simply classifying and labeling dangerous substances and preparations would not always suffice to protect human health. Second, Member States were individually restricting the marketing and use of chemicals in ways that created barriers to trade within the European Communities.

Directive 76/17/69 enabled authorities to restrict the use of specified chemicals throughout Europe. Such restrictions typically meant that the use of the substance and the preparations and products containing it were controlled; for example:

- The substance could only be used for certain applications by professional users, and marketing to and use of the substance by the public were banned.
- Certain applications or uses posing a particular risk were banned.
- Use of a substance in a product was allowed only below a specified threshold concentration.
- Marketing was only allowed when specific labeling and safety requirements were met.

For example, the Limitations Directive prohibited the use of chloro-1-ethylene for one application, as an aerosol propellant. In some cases, notably with respect to PCBs, the Limitations Directive imposed an outright ban. Restrictions were based on the potential to pose a risk to human health, particularly from carcinogenicity or in some cases a potential risk to the environment. Under the initial Directive and multiple amendments, authorities restricted 42 substances or groups of substances, which included a total of approximately 900 compounds.*

* **Note:** The list of substances regulated under Directive 76/769/EEC was later incorporated into Annex XVII of REACH, according to Commission Regulation (EC) No 552/2009 of 22 June 2009 amending Regulation (EC) No 1907/2006 of the European Parliament and of the Council on the Registration, Evaluation, Authorisation and Restriction of Chemicals (REACH) as regards Annex XVII.

3.2.3 Sixth Amendment and Existing Substances Regulation

The Sixth Amendment to Directive 67/548/EEC [78] included the following summation:

> ... in order to control the effects on man and the environment it is advisable that any new substance placed on the market be subjected to a prior study by the manufacturer or importer and a notification to the competent authorities conveying mandatorily certain information ... it is, moreover, important to follow closely the evolution and use of new substances placed on the market, and ... in order to do this it is necessary to institute a system that allows all new substances to be listed ...

The Sixth Amendment distinguished between chemicals existing at the time it was implemented and new chemicals placed on the market after that time as described below.

3.2.3.1 Existing Chemical Substances

Article 13 of the Sixth Amendment simply required the authorities to inventory existing chemicals. Over a decade later, the assessment of those existing chemicals began with the promulgation of the Existing Substances Regulation.

3.2.3.1.1 Article 13 of the Sixth Amendment

Article 13 of Directive 67/548/EEC, as amended by Directive 79/831/EEC, required the authorities to compile a list of all chemical substances existing as of September 1981. The Commission Decision of 11 May 1981 (81/437/EEC) [79] articulated the criteria by which Member States were to supply information to the European Commission for the inventory of existing substances, known as the European Inventory of Existing Commercial Chemical Substances (EINECS). EINECS listed 100,106 chemical substances existing in 1981. Many of these substances were used in relatively small amounts; circa 2001, for example, only approximately 30,000 of those substances were marketed in quantities greater than 1 tonne per year and just under 3,000 were High Production Volume Chemicals (HPVC) manufactured or imported at over 1,000 tonnes per year [80,81]. Although the Sixth Amendment mandated that chemicals be listed, it did not require that any of the existing chemicals be tested or the risks of their use assessed.

3.2.3.1.2 Existing Substances Regulation

In 1993, the European Council took action to require the assessment of existing chemicals. The Existing Substances Regulation (EEC) 793/93 [82] specified a four-step process for evaluating existing chemicals: data collection, priority setting, risk assessment, and, if necessary, risk reduction. The regulation required any company which produced or imported an existing substance in a quantity exceeding 10 tonnes per year to provide data on that substance. The Existing Substances Regulation prioritized data collection as follows:

- *Phase I:* HPVCs, that is, those with a production or import volume above 1,000 tonnes per year, included in Annex I of the Regulation. Data had to be submitted by June 1994.
- *Phase II:* All other substances of a production or import volume in excess of 1,000 tonnes/year. Data had to be submitted by June 1995.
- *Phase III:* Low Production Volume Chemicals (LPVCs) produced or imported between 10 and 1,000 tonnes per year. Data were required by June 1998.

Table 3.5 summarizes the type of information collected [81], which included information about production and use, and also the available data on the behavior of the substance.

This effort amassed information on 2,782 HPVCs and 7,825 LPVCs [83], which was made available to the public in the International Uniform ChemicaL Information Database (IUCLID).

Because submitters were providing only the available data on existing chemicals, rather than data from a prescribed testing regimen, information was not available for every endpoint. Figure 3.3 illustrates the data available for some critical endpoints as a percentage of the IUCLID entries for HPVCs [81]. (The submittals compiled in IUCLID reflected far more data-rich dossiers than the PMN submittals under the US EPA's TSCA. As discussed previously, approximately 44% of PMN submittals have contained some toxicological data [typically on acute toxicity], 9% contained some ecotoxicological data [again, typically on acute toxicity], and 6% contained information on biodegradability. Fifty-four percent contained no test data.)

The authorities identified substances that would undergo risk assessments based on the data submitted by manufacturers and importers and on the national lists of priority substances. Of over 100,000 existing substances, which included approximately 11,000 HPVCs and LPVCs, 141 were

TABLE 3.5

Data Collected Under 1993 Existing Substances Regulation

Data Required	HPVCs	LPVCs
Produced/imported quantities	✓	✓
Classification and labeling under Directive 67/548	✓	✓
Reasonably foreseeable uses	✓	✓
Physicochemical properties	✓	
Chemical fate and pathways	✓	
Toxicological properties	✓	
Ecotoxicological properties	✓	

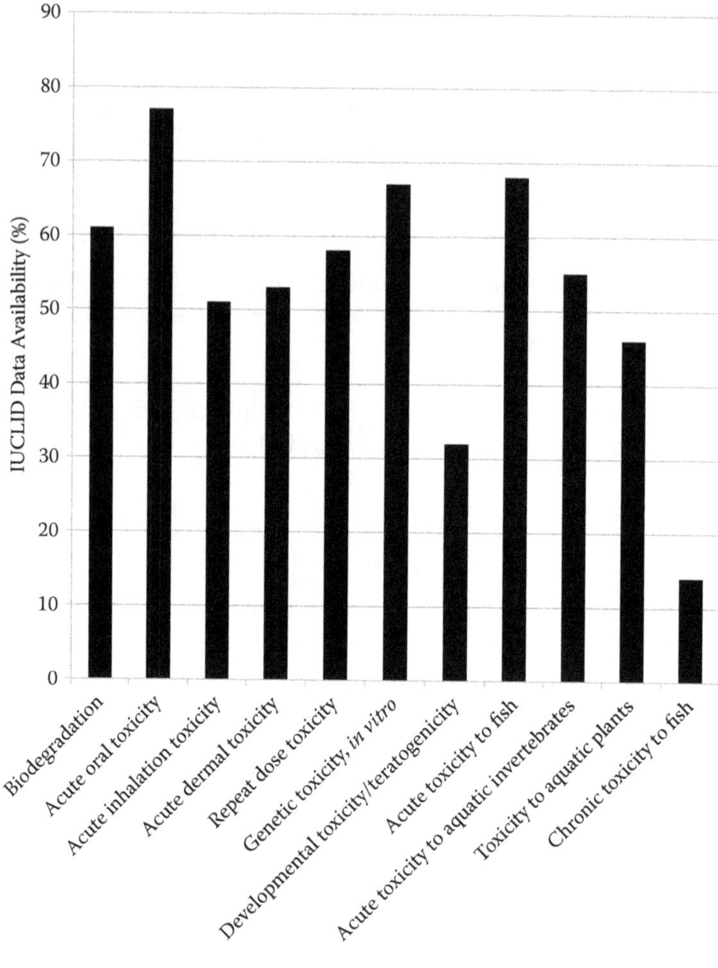

FIGURE 3.3
Available data for High Production Volume Chemicals.

identified for risk assessment under the Existing Substances Regulation and Regulation EEC 1488/94. Depending upon the results of the assessment, the authorities could propose risk reduction measures under the Existing Substances Regulation.

For illustration, consider the risk assessment of toluene, an HPVC, under the Existing Substances Regulation [84] and the outcome of that assessment. The study estimated exposures throughout the life cycle of toluene, considering the production of toluene and the annual use of some 2,750,000 tonnes (circa 1995) of the substance in Europe in applications that ranged from the production of other chemicals (approximately 80% of toluene usage) to serving as a solvent carrier in paints, thinners, adhesives, and inks, and including use as a processing aid in the production of pharmaceutical and other chemical products. The European Union System for the Evaluation of Substances (EUSES) model predicted that these uses would emit 1,090 tonnes per day (t/d) toluene to the air, 180 t/d to wastewater, 77 t/d to surface water, and 2.4 t/day to industrial soil within the European continent. One-third of the total environmental release would result from private use of toluene.

The assessment concluded that specific measures were needed to limit the risks to workers, consumers engaged in spray painting or carpet laying, humans exposed via the environment, the aquatic ecosystem, the terrestrial ecosystem, the atmosphere, and microorganisms in sewage treatment plants.

In light of these conclusions, Denmark made a series of recommendations as the rapporteur to limit risks [85, p. 117]. As evident from the summary which follows, these recommendations included measures under the Limitations Directive and under other environmental laws that controlled emissions to air and water. In short, the rapporteur recommended:

- To consider at Community level marketing and use restrictions in the Limitations Directive for the substance as such or in preparations for use in adhesives and spray paint.
- "In order to limit risks from environmental exposures, that the European Commission should consider the inclusion of toluene in the priority list of Annex X to Directive 2000/60/EC (Water Framework Directive) during the next review of the Annex but that, in the meantime, toluene should be considered as a relevant List II substance in Council Directive 76/464/EEC 35 on pollution caused by certain dangerous substances discharged into the aquatic environment of the Community, thus requiring the establishment of national quality objectives, monitoring and eventual reduction measures, as to ensure that concentrations in surface water systems do not exceed the quality objective" [85].
- To facilitate permitting under Council Directive 96/61/EC (Integrated Pollution Prevention and Control) that toluene be included in the ongoing work to develop guidance on *Best Available Techniques.*
- Local emissions to the environment should, where necessary, be controlled by national rules to ensure no expected risk for the environment.

The recommendations also noted the potential risks to workers who might inhale toluene or contact it with skin or eyes, but concluded that the existing legislation for workers' protection gave an adequate framework to limit the risks of the substance to the extent needed.

The European Communities transmuted these recommendations into a proposed Directive that defined enforceable restrictions on toluene in consumer products. The restrictions reflected both the results of the risk assessment and the regulatory framework of the Existing Substances Regulation. The *Proposal for a Directive of the European Parliament and of the Council* relating to restrictions on the marketing and use of toluene and trichlorobenzene [86] explained the latter in pragmatic terms:

> Due to the fact that certain uses of chemicals cannot be controlled under certain conditions, safety for human health can only be ensured by prohibiting such uses of the substances and preparations concerned.
>
> The aim of the proposal is to preserve the Internal Market. When Member States adopt national provisions restricting the marketing and use of dangerous substances and preparations there will be obstacles to trade because of differences in legislation between Member States. The Draft Proposal aims to improve the conditions for the functioning of the Internal Market to the benefit of the protection of people's health and safety.

The German Federal Environment Agency concluded upon reviewing the proposed rule that [87, pp. 40–41]:

> … [it] bans the use of Toluene only in certain coatings and adhesives, pursuant to the following conditions:
>
> - Spray paint is the only type of coating included in the ban.
> - The restrictions apply to consumer products only and not to professional products, i.e., products that are used for commercial purposes can still contain Toluene (e.g., adhesives used by carpet and flooring installers).
> - In these products, the mass concentration of Toluene is still limited to <0.1% (i.e., 1 kg of product may not contain more than 0.99 g of pure Toluene).
>
> … this ban will not reduce the number of exposure scenarios by a significant amount (i.e., the use of Toluene in spray paints and adhesives). The far more prevalent uses of Toluene are excluded from the … Directive, as are the attendant protective mechanisms for humans and the environment. In short, the proposed ban on the use of Toluene pertains to a portion of the identified risks only.

Despite such concerns, Annex I of Directive 76/769/EEC, the Limitations Directive* [88], ultimately restricted only the following use:

> Toluene CAS No. 108-88-3//May not be placed on the market or used as a substance or constituent of preparations in a concentration equal or higher than 0.1% by mass in adhesives and spray paints intended for sale to the general public.

* This provision, along with the other restrictions imposed by the Limitations Directive, has subsequently been incorporated into Annex XVII of REACH, and the risk-assessment process subsumed into REACH.

In summary, the Existing Substances Regulation, which pertained to 100,106 substances listed on EINICs, drove the risk assessment of 141 substances under the Existing Substances Regulation and Regulation EEC 1488/94. Some of these substances came to be regulated under the Limitations Directive. As the overview of the decision-making process for toluene illustrates, authorities considered potential risks from exposure throughout the product life cycle in determining the appropriate risk-management measures; such measures were not always selected without controversy. Further, the record indicates that the authorities were clearly aware of the trade consequences of chemical restrictions applied unevenly throughout the region.

The corresponding process for assessing new chemical substances is described below.

3.2.3.2 New Chemical Substances

New chemicals had to be tested under the Sixth Amendment to Directive 67/548/EEC. The Sixth Amendment required the manufacturer or importer to submit a dossier containing these data to the member nation-state in which manufacture occurred or into which the substance was imported, and further required the competent authorities in that country to share the information with other member nation-states.

The Sixth Amendment left the details of implementation to Member States. The way in which each Member State implemented the Directive reflected the regulatory milieu and social values of each country [89]. West Germany passed the Chemicals Act of 1980, and France, the Law on the Control of Chemicals, for example. Thus the implementation of the Sixth Amendment reflected a patchwork of country-specific requirements.

Between 1981 and 2007 manufacturers and importers notified the competent authorities of over 5,000 new substances, which were listed on the European List of Notified Chemical Substances (ELINCS). The Sixth Amendment left it to the competent authorities to utilize the testing data to assess the potential risks from these new chemical substances.

The Seventh Amendment, promulgated in 1992, described the requirements for performing risk assessments as follows [90]:

> … the competent authority which received notification shall carry out an assessment of the risks …. The assessment shall include recommendations on the most appropriate method for testing the substance and, where appropriate, also include recommendations on measures which will enable the risk for man and the environment in connection with the marketing of the substance to be lessened.

One such substance was known as Navy Blue (EC #405-665-4), described more formally in ELINCS as the reaction mass of disodium (6-(4-anisidino)-3-sulfonato-2-(3,5-dinitro-2-oxidophenylazo)-1-naphtholato)

(1-(5-chloro-2-oxidophenylazo)-2-naphtholato)chromate(1-) and trisodium bis(6-(4-anisidino)-3-sulfonato-2-(3,5-dinitro-2-oxidophenylazo)-1-naphtholato)chromate(1-). Manufacturers used Navy Blue to dye wool and polyamide fabric and it has often been used to dye automobile upholstery fabric.

As a new chemical substance, Navy Blue was first registered in Switzerland in 1990 and then subsequently in seven other Member States. The total European sales volume at the time of the assessment was estimated at between 9 and 90 tonnes per year [87]. Consequently, as reported by the German Federal Environment Agency [87, p. 47],

> Navy Blue reaches the environment (surface water) via waste water from textile installations as well as numerous diffuse sources stemming from all usage levels (manufacture, formulation, and preparations). Even when used in accordance with the applicable regulations, such contamination is unavoidable owing to the fact that not all of the dye used in textiles is fixed to the fibers. The non-fixed dye, or residual liquor, reaches the environment via waste water treatment plants where it can only be eliminated in part as it is does not degrade readily. Thus, if Navy Blue is used in the recommended concentrations, it will inevitably pollute surface waters and will harm the aquatic organisms therein. Discharge monitoring is probably not a viable solution to this problem since most Navy Blue is generated by [small and medium enterprises], which means that discharge comes from numerous direct and indirect sources that are not amenable to monitoring.
>
> Since Navy Blue is not very volatile, emission to air does not come into play. No data is [sic] available for either the aquatic or terrestrial compartment because under the New Substances Regulation, assessment of these factors is only required for production volumes of upwards of 100 t/year .... The predicted no effect concentration (PNEC) for Navy Blue is estimated to be PNECaqua = 1.6 [micrograms per liter]. However, the predicted environmental concentration (PEC) for Navy Blue is far higher, thus pushing the PEC/PNEC ratio beyond 1 to approximately 1.9. ... [I]n this case all possible ways to minimize environmental discharge of Navy Blue were explored exhaustively with the user by means of additional data that was [sic] provided. But inasmuch as, following the conclusion of this dialogue phase, it was felt that, from a technical standpoint, the risk assessment (PEC/PNEC >1) was unlikely to change ... a total prohibition on the use of Navy Blue has been proposed.

As a consequence of the risk assessment, Navy Blue was classified as very toxic to aquatic organisms with the potential to cause long-term adverse effects in the aquatic environment. It was also classified as a substance that could cause sensitization of humans by skin contact. In 2004 Navy Blue was added to Annex I to Directive 76/769/EEC in the category of azo dyes:

> Azo dyes, which are contained in the "List of azo colourants" that is hereby added to the Annex, may not be placed on the market or used

for colouring textile and leather articles as a substance or constituent of preparations in concentrations higher than 0.1% by mass.

This vignette illustrates one consequence of the gradual evolution of chemical control regulations. Navy Blue was first notified as a new chemical with the submittal of data in 1990. Additional tests were performed after the initial review of the data [87]. The Seventh Amendment formalized risk assessment requirements in 1992. The risks from Navy Blue had been assessed by 1994 [87]. It was another 20 years before the restriction on sales and use occurred.

Notification under the Sixth Amendment (and related risk assessments under the Seventh Amendment as described above for Navy Blue) continued until 1 June 2008. On that date the notification scheme was revoked and replaced by Regulation (EC) No 1907/2006 concerning the Registration, Evaluation, Authorisation and Restriction of Chemicals (REACH). ELINCS then contained 8,433 notifications, representing 5,292 substances in total [91].

The preceding discussion of the Sixth Amendment and the Existing Substances Regulation explains how, of the 105,398 new and existing chemical substances, competent authorities assessed the potential risks from exposure to each of 5,433 substances and ultimately restricted approximately 900 substances under the Limitations Directive. It also illustrates some of the concerns with the legislation that led to calls for reform, and ultimately to the implementation of REACH.

3.2.4 Calls for Reform

The patchwork of laws described above had several shortcomings, among them the limited availability of data for the bulk of chemicals in commerce and the schemes for risk assessment and promulgation of risk management measures [92]. Europa, the official website for the European Union, describes the results of these shortcomings: "[T]he previous system often proved itself to be incapable of identifying risks posed by many chemicals and was slow to act when risks were identified" [93].

Consequently the European Communities determined that new chemical control legislation was needed and defined seven objectives for revised chemical regulation [80]:

- Protection of human health and the environment
- Maintenance and enhancement of the competitiveness of the chemical industry in the European Union
- Prevention of fragmentation of the internal market
- Increased transparency
- Integration with international efforts
- Promotion of non-animal testing

- Conformity with European Union international obligations under the World Trade Organization

REACH, the European Community Regulation on chemicals and their safe use (EC 1907/2006), was the consequence of those deliberations.

3.2.5 REACH

REACH pertains to the Registration, Evaluation, Authorisation and Restriction of Chemicals [94]. It entered into force on 1 June 2007 and is being implemented in stages with registration to be completed by 1 June 2018. It is no exaggeration to say that this regulation revolutionized chemical control laws. First, it required manufacturers and importers to characterize the physicochemical properties and toxicity of nearly every chemical substance in commerce at over 1 metric tonne per year, an unprecedented requirement to generate data. In short, "no data, no market." Second, it upended the paradigm for risk assessment and risk management. In the words of the German Federal Ministry of the Environment [87, p. 8]:

> REACH shifts the moorings of Existing Substances risk reduction from the realm of legislative and administrative policy and strategy to a far greater emphasis on economic stakeholders taking responsibility for their own actions.

Third, REACH embodies the basic premise that the risks presented by any chemical on the market above a *de minimus* quantity must be acceptable. The manufacturer or importer must demonstrate that risk reduction measures can adequately manage the human health and environmental risks from exposure to each chemical. Further, REACH provides mechanisms to restrict or ban chemical substances.

To accomplish its objectives, the regulation runs to over 500 pages of text; the guidance documents compiled by the European Chemicals Agency (ECHA) amount to thousands more. A brief summary of the requirements of REACH follows, beginning with the first step in the regulation of a chemical substance: registration.

3.2.5.1 Registration

REACH requires that manufacturers and importers register all chemical substances manufactured or imported into the European Union in amounts above 1 metric tonne per year. The threshold applies per substance and per manufacturer/importer. Registration requirements can apply to pure chemical substances, substances in mixtures ("preparations"), and certain substances in finished products ("articles").

Certain substances are exempt from registration such as:

- Substances used for research and development
- Waste
- Pesticides and biocides, and other substances already regulated
- Polymers
- Chemical substances notified under previous EU legislation (Directive 67/548/EEC)

For those substances that require notification, the requirements and deadlines depend upon the tonnage in commerce. REACH requires that chemicals in commerce at levels >1,000 tonnes per annum be notified by 2010; 100–1,000 tonnes per annum by 2013, and 1–100 tonnes/annum by 2018.

Central to registration is the Chemical Safety Assessment (CSA), in which the registrant identifies and describes the conditions under which the manufacturing and use of a substance are considered to be safe. A CSA must be performed by registrants for substances manufactured and imported in quantities at or above 10 tonnes per year and by downstream users if their uses are not addressed by their supplier. The CSA entails three major steps, hazard assessment, exposure assessment, and risk characterization, which are documented in the Chemical Safety Report (CSR). Figure 3.4* summarizes the process [95].

The process begins with gathering and assessing all available data. As shown in Table 3.6 (later in the chapter), the amount of data required depends on the annual tonnage manufactured or imported, with the largest amount of data required for the chemicals manufactured or imported at the highest levels. Some of the data listed in Table 3.6 are only appropriate for certain physical states or types of chemicals or chemical behavior, or can be waived under specified circumstances.

If the test data for an endpoint are of poor quality, one can estimate the value using the weight of evidence. If test data are not available at all for a particular endpoint, REACH allows for read-across to the test results for a similar compound within certain restrictions, or estimation using QSAR. If after considering such techniques the data still do not suffice to characterize the substance, then the manufacturer or importer must commission testing. Any plans to test toxicity using vertebrate animals must undergo review and approval by ECHA before implementation. REACH seeks to minimize animal testing. When it is absolutely necessary, REACH requires that minimum pain, distress, and suffering is caused to the animals [96].

After compiling data, the registrant must perform a hazard assessment [97]. (This step corresponds to hazard identification and characterization in the generic risk assessment process described in Chapter 2.) ECHA describes this process as follows [98, p. 5]:

* Source: European Chemicals Agency, http://echa.europa.eu/

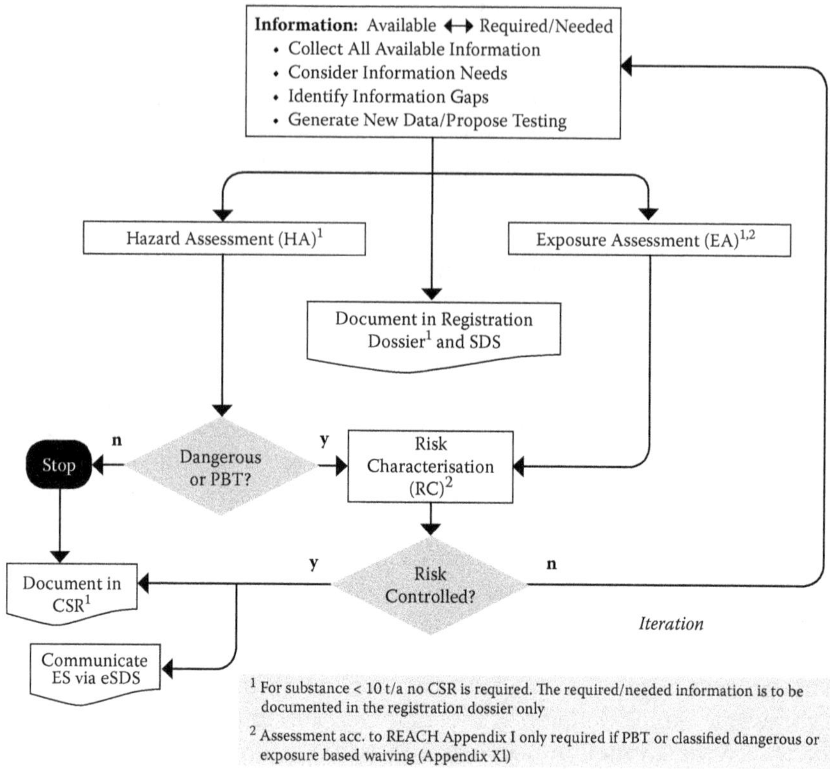

FIGURE 3.4
Overall process for Chemical Safety Assessment.

> The objective of the hazard assessment is to identify the hazards of the substance, assess their potential effects on human health and the environment, and determine, where possible, the threshold levels for exposure considered as safe (the so called no-effect levels).

In other words, the substance under consideration must be evaluated using the formal rules for classifying hazards (i.e., Directive 67/548/EEC) to determine if it is "dangerous." In addition, values for DNELs, DMELs, and PNECs must be derived and the substance assessed to determine whether it would be considered a PBT or very persistent/very bioaccumulative (vPvB) substance based on the data assembled in the first step of the CSA.

REACH Annex XIII specifies the criteria for PBT/vPvB substances. A compound is considered to be PBT if it meets the following criteria:

- Persistence: Half-life greater than the following benchmarks
  - 60 days in marine water, or
  - 40 days in fresh- or estuarine water, or

  - 180 days in marine sediment, or
  - 120 days in fresh- or estuarine sediment, or
  - 120 days in soil
- Bioaccumulative: BCF > 2,000 L/kg
- Toxic:
  - No observed effects concentration (NOEC) < 0.01 mg/L for marine or freshwater organisms, or
  - Substance is classified as carcinogenic (category 1 or 2), mutagenic (category 1 or 2), or toxic for reproduction (category 1, 2 or 3), or
  - There is other evidence of chronic toxicity as indicated by the classifications T, R48, or Xn, R48 according to Directive 67/548/EEC.

A substance is vPvB if it meets the following criteria:

- Persistence: Half-life greater than the following benchmarks:
  - 60 days in marine, fresh- or estuarine water, or
  - 180 days in marine sediment, fresh- or estuarine sediment, or
  - 180 days in soil
- Bioaccumulative: BCF > 5,000 L/kg

If the hazard assessment indicates that the substance is not dangerous and is not PBT/vPvB, then the CSA is complete. However, if the substance meets the criteria for classification as dangerous or is considered a PBT/vPvB, then Article 14(4) of REACH requires that the registrant must perform an exposure assessment and risk characterization. A registrant need only assess its own tonnage; the cumulative mass of multiple registrants is not assessed.

The exposure assessment, in ECHA's words [98, p. 5]:

> ... is the process of measuring or estimating the dose or concentration of the substance to which humans and the environment are or may be exposed, depending on the uses of the substance.
>
> Within the exposure assessment, the definition of the conditions under which the substance is manufactured and used is critical in order to determine the levels of exposure. The information on the conditions under which a substance is manufactured and used is called the exposure scenario under REACH. For each exposure scenario, the exposure levels of humans and the environment need to be determined. The exposure scenarios will cover all identified uses and life stages of the substance.

The registrant(s) must develop exposure scenarios that describe how they and their customers ("downstream users") manufacture and use the substance throughout the product life cycle [99]. Each scenario includes operational conditions (e.g., the duration and frequency of use or the amount used) and necessary risk management measures (such as the use of personnel protective equipment [PPE] or pollution control equipment). Ideally, measured data would be available. More commonly, the registrant uses information on operational conditions and risk-management measures to estimate exposure concentrations for each scenario. Human exposures and environmental exposures are estimated separately.

Table 3.7 lists the models commonly used to predict human exposures under REACH; ECHA also refers exposure assessors to models developed by the US EPA and discussed earlier in this chapter [100–105]. Each of these tools has various features, strengths, and weaknesses which must be understood by an experienced modeler before selecting and applying an exposure model.

Environmental exposure assessments occur at different scales and in different media. Predicted environmental concentrations (PECs) may be developed for air, water, sediment, or soil, at local, regional, or continental scales.

Exposure assessments reflect three fundamental patterns of release:

- Industrial setting scenario at a local scale
- Wide dispersive scenario at a local scale
- Releases in a standard region for all identified uses

As appropriate to the scenario, one may consider release to the air, directly to water, indirectly to water through a sewage treatment plant (STP), or to land. Upon release, the concentration and nature of the substance begins to change as a result of dilution, volatilization, sorption, degradation (biotic or abiotic), and uptake by organisms (with the potential for bioconcentration or biomagnification). The effects of these processes can each be estimated, but modeling efforts can obviously become very complicated due to the number of processes involved. ECHA provides a series of equations and simplifying assumptions that can be used to develop a first approximation [106]. Additionally, analysts use the EUSES model.

The EUSES model was developed to support the risk assessment of chemicals under various regulations. EUSES is a spreadsheet-based tool that incorporates SimpleBox, a multimedia fate model, and SimpleTreat, which simulates the distribution and elimination of chemicals in sewage treatment plants. The model estimates the concentrations of a substance in air, water, soil, and sediment at local and regional scales. It simulates the steady-state transport of chemicals between media and scales, and removal of the chemical by degradation and some "disappearance processes" (e.g., leaching to the groundwater, which is not then modeled) [106,107].

In summary, the exposure assessment provides estimates of the concentrations of the chemical under consideration in various media in different settings that may be encountered by people or by ecological receptors. These

TABLE 3.6
Data Requirements under REACH Based on Annual Tonnage

Study	Annex VII 1–10 t	Annex VIII 10–100 t	Annex IX 100 –1,000 t	Annex X ≥1,000 t
Physicochemical Properties				
State of Substance	✓	✓	✓	✓
Melting/Freezing Point	✓	✓	✓	✓
Boiling Point	✓	✓	✓	✓
Relative Density	✓	✓	✓	✓
Vapor Pressure	✓	✓	✓	✓
Surface Tension	✓	✓	✓	✓
Water Solubility	✓	✓	✓	✓
Partition Coefficient *n*-Octanol/Water	✓	✓	✓	✓
Flash Point	✓	✓	✓	✓
Flammability	✓	✓	✓	✓
Explosive Properties	✓	✓	✓	✓
Self-Ignition Temperature	✓	✓	✓	✓
Oxidizing Properties	✓	✓	✓	✓
Granulometry	✓	✓	✓	✓
Stability in Organic Solvents and Identity of Relevant Degradation Products			✓	✓
Dissociation Constant			✓	✓
Viscosity			✓	✓
Toxicological Information				
Skin Irritation	✓	✓ (*in vivo*)	✓ (*in vivo*)	✓ (*in vivo*)
Eye Irritation	✓	✓ (*in vivo*)	✓ (*in vivo*)	✓ (*in vivo*)
Skin Sensitization	✓	✓	✓ (*in vivo*)	✓ (*in vivo*)
Mutagenicity	✓	✓	✓ (+*in vivo*)	✓ (+*in vivo*)
Acute Oral Toxicity	✓	✓	✓	✓
Acute Dermal Toxicity		✓	✓	✓
Acute Inhalation Toxicity		✓	✓	✓
Repeat Dose Oral Toxicity (28-day)		✓	✓	✓
(>12 months if warranted)				
Sub-Chronic Toxicity (90-Day)			✓	✓
Reproductive/Developmental Screening		✓		
Reproductive/ 2-Generation Reproductive Testing			✓ if needed	✓ if needed

Continued

TABLE 3.6 (*Continued*)

Data Requirements under REACH Based on Annual Tonnage

Study	Annex VII 1–10 t	Annex VIII 10–100 t	Annex IX 100 –1,000 t	Annex X ≥1,000 t
Developmental			✓ (prenatal)	✓
Toxicokinetics		✓ (if available)	✓ (if available)	✓ (if available)
Carcinogenicity				✓
Environmental Fate and Ecotoxicological Information				
Short-Term Invertebrate (*Daphnia*) Toxicity	✓	✓	✓	✓
Algal Growth Inhibition	✓	✓	✓	✓
Short-Term Fish Toxicity		✓	✓	✓
Activated Sludge Respiration Inhibition		✓	✓	✓
Daphnia Long-Term Toxicity			✓	✓
Long-Term Toxicity Testing on Fish			✓	✓
Bioaccumulation in Fish			✓	✓
Degradation (Biotic)	✓	✓	✓	✓
Abiotic Degradation		✓	✓	✓
Hydrolysis as Function of pH		✓	✓	✓
Identification of Degradation Products			✓	✓
Soil Adsorption/Desorption Screening Test			✓	✓
Effects on Terrestrial Organisms: Short-term toxicity to invertebrates; effects on soil organisms; short-term toxicity to plants			✓	✓
Long-term Effects on Terrestrial Organisms: invertebrates; plants; soil organisms reproductive toxicity to birds				✓

concentrations form the basis for risk assessment, which is the next step in the process.

ECHA summarizes the process of risk assessment and its iterative nature as follows [98, p. 14]:

> … an initial risk characterisation can be performed by comparing the expected levels of exposure to the predicted no effect levels from the hazard assessment. If as a result of the initial risk characterisation, risks can be demonstrated to be under control, the initial exposure scenario(s) will become the final exposure scenario(s). If risks are not controlled, further refinement of the CSA will be required, until the safe use of the substance can be proved or its use or uses advised against. The refinement involves one or more iterations of the CSA.

TABLE 3.7

Models Used to Predict Human Exposures under REACH

Receptor	Endpoint	Tier	Model
Worker	Inhalation Dermal	1	ECETOC TRA tool (based on EASE, or Estimation and Assessment of Substance Exposure, model). This tool predicts inhalation exposure by analogy with similar situations in which the exposure concentration has been measured, and estimates dermal exposure based on physical state, pattern of use, and pattern of control.
	Inhalation	1	Easy-to-Use Workplace Control Scheme (COSHH-BAuA tool)/ EMKG-Expo-Tool. Screening-level tool that predicts exposure via inhalation considering two principal factors: the exposure potential of the handled substance and the applied control strategy.
	Inhalation	1	Stoffenmanager exposure model. Estimates median task-based exposure level from statistical analyses of the relation between deterministic scores and around 1,000 real exposure measurements.
	Inhalation	2	Advanced REACH Tool (ART). Produces exposure estimates by combining a mechanistic model and empirical data from an exposure database using a Bayesian statistical process.
	Dermal	2	RISKOFDERM. Calculates potential exposure per minute (for hands and/or remainder of the body), assuming no PPE, for six dermal exposure operational (DEO) units: handling contaminated objects, manual dispersion, hand tool dispersion, spray dispersion, immersion, and mechanical treatment.
Consumer	Oral Inhalation Dermal	1	ECETOC TRA Consumer tool. Includes default values for 46 specific product and article types relevant for consumer use to enable screening-level calculations of potential exposure. Default values are consistent with ConsExpo.
	Oral Inhalation Dermal	2	ConsExpo 4.1. Based on exposure-relevant information for main categories of consumer products (children's toys, cleaning products, cosmetics, disinfectant products, do-it-yourself products, paint products, pest control products), estimates indoor exposure by adults and children.
		2	GExFRAME. A generic consumer exposure modeling system intended to harmonize source models, fate and transport models, and exposure/risk models, GExFrame is being developed to accommodate various algorithms including those that reside in CONSEXPO, E-FAST, CARES, and similar models.

There are basically three options to refine the CSA process:

- improve the hazard assessment by obtaining more data;
- improve the exposure assessment by ensuring that the exposure estimation is realistic and reflects the conditions of use defined in the initial exposure scenario, or
- improve the conditions of manufacturing or use, e.g. by introducing more stringent risk management measures or changing the operational conditions in the exposure scenario.

Thus the result of the risk assessment may be to re-examine the underlying data and calculations and to invest additional resources in refining the input to the risk assessment.

In order to bring any chemical to market, the CSA must demonstrate that for the specified exposure scenario(s), risks are adequately controlled. In ECHA's words [98, p. 19], this means:

- The likelihood and severity of an event occurring due to the physicochemical properties of the substance is negligible.
- The estimated exposure levels do not exceed the appropriate DNEL/DMEL or PNEC.
- For substances for which a DNEL/DMEL or PNEC cannot be determined, the emissions and exposures are minimised by the implementation of the exposure scenario to the level that they do not pose risk.

When these standards are met, the registrant proceeds to document the conclusions of the CSA in a CSR which is submitted to ECHA. The final exposure scenario(s), which define the operational conditions and risk-management measures required to ensure the safe use of the substance, are communicated to downstream users through the Safety Data Sheet (SDS).

3.2.5.2 Evaluation

Evaluation can take two forms under REACH. First, ECHA performs a compliance check on 5% of the dossiers submitted to register chemicals. This evaluation considers the completeness and quality of the information in the dossiers. ECHA also evaluates dossiers that call for vertebrate testing, as noted above.

Second, REACH requires that selected substances be evaluated in depth. ECHA and the Member States identify substances suspected of posing a risk to human health or the environment considering the following criteria developed under Article 44 of REACH [108]:

- Hazard:
    - Known or suspected PBTs/vPvBs

    - Known or suspected endocrine disruptors
    - Known or suspected carcinogens, mutagens, or reproductive toxicants (CMRs)
    - Known or suspected sensitisers
- Exposure-related selection criteria:
    - Wide dispersive use
    - Number of using sites if emission due to industrial use
    - Consumer use and exposure of sensitive subpopulations such as children
    - Aggregated tonnage
- Risk-related selection criteria:
    - The CSR indicates that the risk characterization ratio is not far below 1 (for human and/or environmental exposure)
    - Cumulative exposure from structurally related substances with critical hazardous properties (e.g., similar endocrine disrupting property such as antiandrogenic or estrogen-like effect)

Substances of particular concern are included in the Community Rolling Action Plan, or CoRAP. Once a substance is on the CoRAP, the Member State performing the review has one year to determine that the risk is in fact acceptable, or else that the registrants must provide more information to characterize a potential risk. If the latter, then submittal of an updated registration dossier triggers a second one-year review period. This second evaluation may find that the additional information demonstrates that manufacture and use of the substance do not present a significant risk. Alternatively, the Member State may propose risk-management measures such as occupational exposure limits or authorization or restriction procedures [109].

3.2.5.3 Authorization

Authorization refers to a multistep process that can culminate in the designation of a substance as one which requires authorization from ECHA for use. The process begins when ECHA or a Member State proposes that a substance be considered as a Substance of Very High Concern (SVHC)* based on a so-called Annex XV SVHC dossier. Member States nominate a compound as an SVHC when it is known to meet criteria specified in Article 57 of REACH, that is, substances classified as CMR, PBT, vPvB, and

* The protocol for naming a substance an SVHC should not be confused with the CoRAP evaluation. The CoRAP provides the authorities with an opportunity to gather more information on a substance that they suspect may be of concern under Article 44 and to evaluate it further. In contrast, a substance is nominated as an SVHC when it is known to be of concern according to Article 57.

substances for which there is scientific evidence of probable serious effects (such as endocrine-disrupting properties) to human health or the environment that give rise to an equivalent level of concern. A full Annex XV SVHC dossier contains not only data on those potential hazards, but also information on the tonnage of the substance on the EU market, the uses of the substance, the resulting release and exposure, and possible alternatives to use of the substance [110].

Once a substance has been nominated as an SVHC, the dossier is made available for public comment and discussion. If no comments are received, the substance is included in the Candidate List. If comments are received, the original proposal and the comments are forwarded to the Member State Committee (MSC) for evaluation. If the MSC does not agree unanimously, the committee refers the matter to the European Communities.

If the authorities ultimately concur with the conclusions of the Annex XV SVHC dossier, the substance is included on the Candidate List. Inclusion on the Candidate List has two effects.

First, the sale of an SVHC on the Candidate List triggers certain communication requirements [111]. Suppliers of pure SVHCs must provide their customers with an SDS, while suppliers of mixtures that contain more than 0.1 wt% of any SVHC must provide an SDS to customers on request. Additional communication requirements pertain to articles.* Manufacturers or importers of articles containing more than 0.1% by weight of any SVHC must provide their customers with adequate information on the safe use and disposal of the article, including the name of the SVHC(s) concerned. Manufacturers and importers of articles must also notify ECHA of the quantities of SVHCs used in their articles if the substance is present in those articles in quantities totaling over one tonne per producer or importer per year and if the substance is present in those articles above a concentration of 0.1 wt%.

Second, ECHA may eventually place a Candidate List substance on the Authorisation List. Twice each year ECHA must recommend to the European Communities a list of priority substances for inclusion in Annex XIV. After considering the opinion of the MSC and public comments, the European Communities decides which of the recommended substances to include in Annex XIV.

If, as a result of this process, an SVHC is included in the Authorisation List (and thus in Annex XIV of REACH), it cannot be placed on the market or used unless that use is exempted or the manufacturer or user applies for and receives authorisation for a specific use. As described in Article 55 of REACH:

> The aim ... is to ensure the good functioning of the internal market while assuring that the risks from substances of very high concern are

* Article 3 of REACH defines the term *article* as "an object which during production is given a specific shape, surface or design which determines its function to a greater degree than does its chemical composition."

> properly controlled and that these substances are progressively replaced by suitable alternative substances or technologies where these are economically and technically viable. To this end all manufacturers, importers and downstream users applying for authorisation shall analyse the availability of alternatives and consider their risks, and the technical and economic feasibility of substitution.

The application for authorization entails a series of steps that begin with a notification of the intent to apply for authorization, ideally eight months in advance of the application. The application itself comprises multiple parts [104]. It must include information on the substance identity, the uses applied for, and the CSR (unless already submitted). It may also include an analysis of alternatives, a substitution plan, socioeconomic analysis, or a justification for not considering certain types of risk.

After submittal, the application undergoes review and public comment in a process anticipated to require over 17 months to complete [112]. Authorization can be granted on two bases:

- *Adequate control route.* The risk to human health or the environment from the use of the substance arising from the intrinsic properties specified in Annex XIV is adequately controlled in accordance with Section 6.4 of Annex I (Art. 60(2)) and taking into account Article 60(3).
- *Socioeconomic assessment (SEA) route.* It can be demonstrated that socioeconomic benefits from the use of the substance outweigh the risk to human health or the environment from the use of the substance and no suitable alternative substances or technologies exist (Article 60(4)). The SEA route is the only possible path to authorization in several specific cases, that is, when the uses cannot be adequately controlled to prevent an unacceptable risk, or the substance presents a human health risk without an established threshold dose, or it is PBT/vPvB, or it presents an equivalent concern.

Given the central questions of this book, the SEA path is particularly interesting. It explicitly allows for the use of a chemical substance when such use could present a risk to human health or the environment. It is too soon to tell how often this path will be followed to authorization or the consequences. The first applications for authorization under REACH were submitted to ECHA in 2013, with comment periods closing in 2014 [113]. All eight applications submitted in 2013, for phthalate compounds, indicated that the risks were adequately controlled. Two of the applications made strong supporting arguments that profound socioeconomic effects would occur from ceasing use but did not solely rely on the SEA route.

If a company receives authorization for an SVHC compound, the ECHA will specify in the authorization the conditions and obligate the holder of the authorization to ensure that exposure is reduced to as low a level as is technically and practically possible. The holder of the authorization must also notify downstream users of the authorization.

3.2.5.4 Restriction

If a chemical poses a broad and unacceptable risk in the opinion of the authorities, its use can be restricted throughout the European Union under this provision of REACH. The 10 steps of the restriction process can take approximately three years [114,115].

1. *Notification of intention to submit a proposal to restrict a chemical.* The restriction process begins when a Member State or ECHA (on request of the European Communities) files a notification of intent due to concerns that a substance poses an unacceptable risk to human health or the environment. The proposal is published in the registry of intentions.
2. *Submission of the proposal.* The notifier submits a formal proposal for restriction, known as an Annex XV dossier, within 12 months after the notification. The submittal explains the need for the proposed restrictions based upon the identified risks and information on alternatives to the substance. The dossier must also present the costs and benefits of the restriction.
3. *Checking of the proposal's conformity.* ECHA checks to see if the submittal conforms to Annex XV. If so, it is released for public comment.
4. *Public consultation.* This six-month period allows for public comment on the proposed restrictions.
5. *Advice on enforceability by Forum.* The Forum of enforcement authorities from the Member States offers an opinion on the enforceability of the proposed restriction.
6. *Opinion by the ECHA Risk Assessment Committee (RAC) and draft opinion by the Committee for Socio-Economic Analysis (SEAC).* These opinions must be provided within nine months of the start of the public consultation. The RAC assesses whether the proposed restriction would appropriately reduce the risk to human health or the environment, considering the dossier, public comment on the dossier, and advice from the Forum. SEAC opines on the socioeconomic impacts of the suggested restrictions, similarly considering the dossier, comments on the dossier, and the Forum's opinion. SEAC offers a draft opinion for public comment as described below.
7. *Public consultation on the SEAC draft opinion.* This consultation period is two months long.
8. *Opinion by SEAC.* Considering the outcome of the public consultation on the draft opinion, SEAC must provide a final opinion on

the socioeconomic impacts of the suggested restrictions within 12 months of the start of the first public consultation on the notification proposal.

9. *Draft amendment of Annex XVII by the European Commission.* Upon considering the identified risks and the costs and benefits of the proposed restriction, the Commission will provide a draft amendment of the list of restrictions in Annex XVII of REACH. This must be produced within three months of receipt of the RAC and SEAC opinions.
10. *European Union decision.* The final decision reflects the input of the Member States and the European Parliament.

If a substance is restricted under REACH it will be added to Annex XVII, which originated with the Limitations Directive.

The proposed restriction on 1,4-dichlorobenzene (1,4-DCB) illustrates the process. Concerns over the potential toxicity of this compound first surfaced under the Dangerous Substances Directive. The European Commission subsequently assessed the risks from exposure to 1,4-DCB under the Existing Substances Regulation, beginning in the mid-1990s.

The 2004 report on the risk assessment noted that approximately 22,950 tonnes of 1,4-DCB were used per year in the European Union in 1985; 20,500 tonnes in 1987; and 16,400 tonnes in 1991. This substance had four primary uses: 49.3% of the total was used as a chemical intermediate (e.g., in making 1,4-dichloro-2-nitrobenzene, a precursor for dyes and pigments); 21.9% in toilet blocks/air fresheners; 28.1% in moth repellants; and 0.7% as a processing aid in the production of grinding wheels [116, p. vii].

The risk assessment determined that based upon the volatility and use of the substance, most of the 1,4-DCB used would be released to the environment. Once in the environment, 1,4-DCB was not expected to persist long: the half-life of 1,4-DCB was estimated to be between 33 and 50 days, and in water 1,4-DCB was readily biodegradable. Based in part on these data the risk assessment concluded with respect to potential risks to the environment that '[t]here is at present no need for further information and/or testing and for risk reduction measures beyond those which are being applied already.' However, toxicity testing showed the potential for human exposure to 1,4-DCB to cause general systemic toxicity, carcinogenicity, and reproductive toxicity. The risk assessment determined that the risks to workers manufacturing 1,4-DCB or products made with that chemical needed to be limited but concluded that existing regulatory controls would suffice to impose those limits.

Concerns regarding consumers' exposure to 1,4-DCB were higher. The substance was used in several everyday applications familiar to many readers: moth repellants for woolen fabrics, and bathroom deodorizers and cleaning products. The European Communities concluded that [117, pp. 21–22]:

> … there is a need for specific measures to limit the risks. This conclusion is reached because of:
>
> - concerns for carcinogenicity as a consequence of inhalation exposure arising from the use of moth repellants, air fresheners, and toilet blocks….
>
> It is recommended:
>
> - to consider at Community Level marketing and use restrictions in Council Directive 76/769/EEC for the use of dichlorobenzene in air fresheners, moth repellants, and toilet blocks.

While stakeholders evaluated the results of the risk assessment, the European Union promulgated REACH. Momentum built to control the use of 1,4-DCB under the Existing Substances regulation had to be redirected under the new regulatory paradigm.

The first action taken to limit the use of 1,4-DCB in the European Union was a decision to restrict its use as a moth repellant under Directive 98/8/EC on biocidal products. Then in 2011, the European Commission requested that ECHA prepare an Annex XV report on the use of 1,4-DCB in air fresheners and toilet blocks [118].

The use of 1,4-DCB in the European Union had shifted since the original risk assessment. By the time the European Union began to consider restriction of 1,4-DCB under REACH, EU manufacturers produced approximately 30,000 tonnes of 1,4-DCB per year. Most of this mass was used as a monomer for polymer production; the remainder was used as a chemical intermediate, in grinding wheels, toilet blocks/air fresheners, and as a laboratory chemical. Only about 800 tonnes per year, including imports, were used in air fresheners and toilet blocks [118]. The amount used in consumer products had decreased significantly since 1985.

ECHA's Annex XV dossier proposed that 1,4-DCB should not be placed on the market or used in toilet blocks or air fresheners to be used in toilets or other domestic or public indoor areas or offices. Exposure could possibly cause some extra cancer cases due to the mitogenic properties of 1,4-DCB; although carcinogenicity has traditionally been thought of as a nonthreshold effect in risk characterization, ECHA concluded that 1,4-DCB had a threshold effect. The RAC, which reviewed the recommendation to determine whether it would appropriately reduce the risk to human health or the environment, concluded that 1,4-DCB could be used safely in products used as air fresheners or to deodorize toilets, homes, offices, and other indoor public areas as long as the concentration was below 1% by weight. The SEAC considered the socioeconomic impacts of the suggested restrictions and concluded that they would not be significant, particularly in light of the decreased use of 1,4-DCB in toilet blocks or air fresheners [118].

In September 2013, the EC published the Draft Commission Regulation amending Annex XVII to Regulation (EC) No 1907/2006 of the European Parliament and of the Council on the Registration, Evaluation, Authorisation and Restriction of Chemicals (REACH) as regards 1,4-dichlorobenzene and notified the World Trade Organisation of the impending regulation [119].

In January 2014 [120], REACH Annex XVII was amended to include the provision that 1,4-DCB "[s]hall not be placed on the market or used, as a substance or as a constituent of mixtures in a concentration equal to or greater than 1% by weight, where the substance or the mixture is placed on the market for use or used as an air freshener or deodoriser in toilets, homes, offices or other indoor public areas." The restriction entered into force in 2015, some 20 years after the initial action was taken to address risks under the Existing Substances directive.

3.2.6 Recapitulation

- Regulation of chemicals in the European Union, previous national laws in the Member States notwithstanding, began with the Dangerous Substances Directive which initially required only appropriate packaging and labeling of dangerous substances.
- The Limitations Directive of 1976 allowed authorities to restrict substances for particular uses. The original Directive restricted only two substances, but the list of regulated substances grew to approximately 900 before this Directive was subsumed into REACH in 2008.
- The Sixth Amendment to the Dangerous Substances Directive required testing of new chemical substances and the exchange of information on new chemicals between nation-states. It did not require the testing of the approximately 100,000 substances already existing in 1981.
- The Seventh Amendment, promulgated in 1992, described the requirements for performing risk assessments of new chemicals.
- The Existing Substances Regulation passed in 1993 initiated the assessment of existing chemicals. This effort compiled existing information on nearly 3,000 high production volume chemicals and almost 8,000 low production volume chemicals. There were 141 substances identified for risk assessment. Some of these substances were regulated under the Limitations Directive.
- REACH, implemented in phases between 2007 and 2018, comprises four parts:
  - *Registration*. REACH will ultimately require the testing and registration of all existing and new chemicals manufactured or imported at greater than 1 tonne per year. The amount of testing depends on the tonnage in commerce. Chemicals which are dangerous or PBT/vPvB must undergo a CSA. If that assessment does not demonstrate that risks can be adequately controlled, the chemical may not be manufactured or imported.
  - *Evaluation*. ECHA audits a fraction of all registration submittals and also evaluates certain substances in depth if they are

identified as potentially posing a notable hazard, may cause widespread or unacceptable cumulative exposure, or may present a nearly unacceptable level of risk.

- *Authorization.* This multistep assessment of SVHC can culminate in a requirement that ECHA specifically authorize any use.
- *Restriction.* If a chemical poses a broad and unacceptable risk in the opinion of the authorities, then its use can be restricted throughout the European Union. Restricted chemicals are added to Annex XVII of REACH, which originated with the substances restricted under the Limitations Directive.

3.3 China

The very first environmental laws in China on the use of natural resources reportedly date back to the Qin Dynasty (221–206 BC). The modern era of environmental regulation began in the 1970s, initially with a law pertaining to marine pollution and expanding through the 1980s and 1990s to include laws governing surface water, air, solid waste, and noise pollution. Laws pertaining to chemical-containing products began to emerge in the 1990s and gained strength two decades later [121–123]. China now regulates chemicals through laws pertaining to imports and exports, and to manufacture and use. Most of these laws and associated regulations came into effect decades after their counterparts in the west. As in other countries, regulations in the People's Republic of China distinguish between chemicals in commerce when regulations came into effect and chemicals newly manufactured after that time.

3.3.1 Existing Chemical Substances

The Chemical Registration Center of the State Environmental Protection Administration (CRC-SEPA, later the Chemical Registration Center of the Ministry of Environmental Protection, MEP) began to inventory existing chemical substances manufactured in or imported into the People's Republic of China in 1992. The first edition of the Inventory of Existing Chemical Substances in China (IECSC) included chemical substances that were manufactured, processed, or used inside China, or imported from outside China for commercial purposes during the period from January 1, 1992 to December 31, 1994 [124]. CRC-SEPA stated the purpose of the inventory as follows [124, p. 3]:

> The "Inventory of Existing Chemical Substances in China" is a technical supportive document for laws and regulations on the prevention and control of pollution caused by chemical substances.

At that time, chemical controls focused on imports and exports of chemicals and on the management of certain hazardous chemicals.

3.3.1.1 Import and Export

The 1994 *Regulations for Environmental Management on the First Import of Chemicals and the Import and Export of Toxic Chemicals* reflected China's commitment to the voluntary predecessor to the Rotterdam Convention and subsequently to the Convention itself. Anyone importing or exporting a designated toxic chemical needed to register with SEPA and obtain a permit. These regulations were amended in 2009 by the *Notice of Registration of the Import and Export of Toxic Chemicals in China* (Circular 113) and again in 2010, and then replaced by Ministry of Environmental Protection (MEP) Order 22, *The Measures for Environmental Administration Registration of Hazardous Chemicals*, in 2012 [125]. Additional requirements were promulgated in the *Detailed Rules on Implementing the Registration of the Environmental Management on the First Import of Chemicals and the Import and Export of Toxic Chemicals* [126].

3.3.1.2 Prohibited Chemicals

Beginning in 2005, the People's Republic of China prohibited certain existing chemicals listed under the Rotterdam Convention on the Prior Informed Consent Procedure for Certain Hazardous Chemicals Pesticides in International Trade and the Stockholm Convention on Persistent Organic Pollutants. The List of Chemicals Banned or Severely Restricted in the People's Republic of China includes primarily (but not exclusively) inorganic compounds and chlorinated organic compounds; as of January 2014, they totaled 162 substances [127,128]. In order to import products containing one or more of these toxic chemicals, a company must obtain a Registration Certificate for Environmental Management on Import/Export of Toxic Chemicals and the Clearance Notification for Environmental Management on Import/Export of Toxic Chemicals [129]. Beginning in 2013, companies that manufacture or use hazardous chemicals must submit an environmental administration registration to environmental protection authorities.

3.3.1.3 Hazardous Chemicals

State Council Decree 591, *Regulations on the Safe Management of Hazardous Chemicals* (originally published in 2002 and amended in 2011), provides another form of chemical control in the People's Republic of China [130,131]. As the name indicates, these regulations apply only to designated hazardous chemicals, which are defined as chemicals that are toxic, corrosive, explosive, flammable, or combustion-supporting and can do harm to people, facilities, or the environment. Importers and manufacturers must register their hazardous chemicals, and manufacturers and vendors have to obtain licenses for manufacturing and sales, respectively. Hazardous chemicals must also be appropriately classified and labeled.

In 2012 the State Administration of Worker Safety (SAWS) issued Order No. 53, *Measures for the Administration of Registration of Hazardous Chemicals*, to specify the detailed requirements for registration of hazardous chemicals under Decree 591. MEP Order No. 22, *Measures for the Environmental Management of Hazardous Chemicals Registration*, was also issued in 2012. It requires companies that produce, use, import, or export hazardous chemicals to register with the government. Further, companies that produce hazardous chemicals of high environmental concern (i.e., those in the Catalog of Priority Hazardous Chemicals for Environment Management) must conduct an environmental risk assessment and report information on emissions to MEP [132].

The Catalog of Hazardous Chemicals, per updates proposed in late 2013 and pending as of late 2014, contains approximately 3,000 chemicals [133,134] including:

- List of dangerous goods/chemicals, which is related to those listed in the United Nation's Recommendations on the Transport of Dangerous Goods
- Catalogue of highly toxic chemicals (approximately 160 substances)
- Priority Hazardous Chemicals (84 substances)
- Other chemicals as determined by authorities

3.3.1.4 Life Cycle Considerations

Regulations on the broad domestic production and use of chemicals arguably began with the *Cleaner Production Promotion Law* of 2002, which provided a bare framework for reducing environmental impacts throughout a product's life cycle. The law requires that, for example, "[i]n the course of technological updating, enterprises shall adopt the following measures for cleaner production [including] replacing highly toxic and harmful raw materials with nontoxic and harmless or less toxic and harmful ones" [121–123]. The effects of this law on chemical usage and substitution are not clear; the initial focus and accomplishments under this law appear to pertain to reductions in resource use and waste generation [135,136].

3.3.1.5 Five-Year Plan Concerning Prevention and Control of Chemical Environmental Risks

The *Twelfth Five Year Plan Concerning the Prevention and Control of Chemical Environmental Risks* [137], published in 2013, may stimulate additional chemical control regulations or policies. The plan itself says (in translation) that the "risk prevention and control situation is becoming increasingly grim." As one account of that 2013 plan relates [138],

> Some toxic and hazardous chemicals that have been phased out or restricted in developed countries are still in scaled production and use in our country. There is clear fact that production capacity of some chemicals with high environmental risk is moving into our country. The Plan identifies 58 kinds of chemicals in three categories as the major control targets for prevention and control of environmental risks in the"12th Five-Year Plan" period based on the sources and types of environmental risks. Six industries including petroleum refinery, coking and nuclear fuel processing, chemical raw materials and products, pharmaceuticals, chemical fiber, smelting and pressing of non-ferrous metals and textile as well as new coal chemical enterprises engaged in manufacturing of oil, natural gas, alkene, demethyl [*sic*] ether and ethylene glycol with coal as raw material will be the major industries and enterprises for prevention and control of environmental risks of chemicals in the "12th Five-Year Plan" period.

The MEP plans to publish a list of restricted substances and a list of substances to be phased out by the end of 2015.

3.3.2 New Chemical Substances

Chemical regulation akin to TSCA or REACH began in 2003, when SEPA passed the *Provisions on the Environmental Administration of New Chemical Substances* (SEPA Order Number 17). The regulatory program changed significantly in 2010 with the issuance of Order Number 7.

3.3.2.1 Provisions on Environmental Administration of New Chemical Substances (Order Number 17)

The *Provisions on the Environmental Administration of New Chemical Substances* (SEPA Order Number 17) [139] required manufacturers and importers to notify and register new chemical substances (defined as substances coming into commerce after 2003) with the CRC-MEP, unless the substance could be exempted based on quantity or specific use.

The Provisions specified that notification must include "the name(s) and the molecular structure; the method(s) applied in measuring; the use(s) of the new chemical substance; the scheduled annual amount of manufacture or import; the physi-chemical [*sic*] properties, toxicological and eco-toxicological characteristics; the measures for accident prevention and emergency responses; and the measures for pollution prevention and elimination and for waste disposal, etc." The form of notification depended upon the amount of the chemical in commerce: basic level (<1 ton per annum), level 1 (10 to 1,000 tons per annum), and level 2 (≥1,000 tons per annum). In addition, the provisions allowed for certain special cases, for example, simplified notification for a substance already listed in four or more existing inventories around the world. The amount of data required increased with the tonnage level [140,141].

The Provisions provided for the authorities to review each dossier for administrative and technical content and then to approve or reject the notification. The technical assessment followed the protocol in *Guidance for the Hazard Assessment of New Chemical Substances* (HJ/T 154-2004), which reportedly resulted in a new chemical being classified according to four health hazard levels (low to extremely high) and five environmental hazard levels (nonhazardous to extremely high), with each evaluation based on the toxicity of the substance and potential exposure. Technical reviewers made four types of recommendations based on this assessment:

- *Extremely high hazard substances:* Prohibit production, sales, and usage
- *High hazard substances:* Limit production, sales, and usage, with clear recommendations for use
- *Medium hazard:* Approve production, sales, and usage, with clear requirements for proper protections
- *Non/low hazard substances:* Approve without any special requirements

The IECSC, which as noted above originated in December 1994, is periodically updated to include new chemical substances. Information on the number of substances on the Inventory over time, and specifically at the time the Provisions took effect in 2003, is difficult to find. As of 2007, it reportedly contained "over 45,000" substances [141]. By 2010, the IECSC contained 45,602 chemical substances in total, including 3,166 confidential substances [142].

An official of the CRC-MEP presented information in 2011 on the numbers of exemptions and notifications accepted between October 15, 2003 and October 14, 2010. These data, shown in Figure 3.5, suggest that relatively few new chemical substances were notified under SEPA Order Number 17 and added to the IECSC [143].

This notification system had limitations; among them according to the CRC-MEP [135] were the following:

- Environmental management was based on hazard assessment instead of risk assessment or prevention.
- Regulation of registered substances only required reporting and did not support the use of risk control measures or supervision of implementation of any restrictions.

3.3.2.2 Revised Provisions on Environmental Administration of New Chemical Substances (Order Number 7)

To strengthen the *Provisions on the Environmental Administration of New Chemical Substances*, the MEP issued Order Number 7 in 2010 [144]. Sometimes referred to as "China REACH," the revised Provisions include additional administrative requirements, reporting requirements, and data collection.

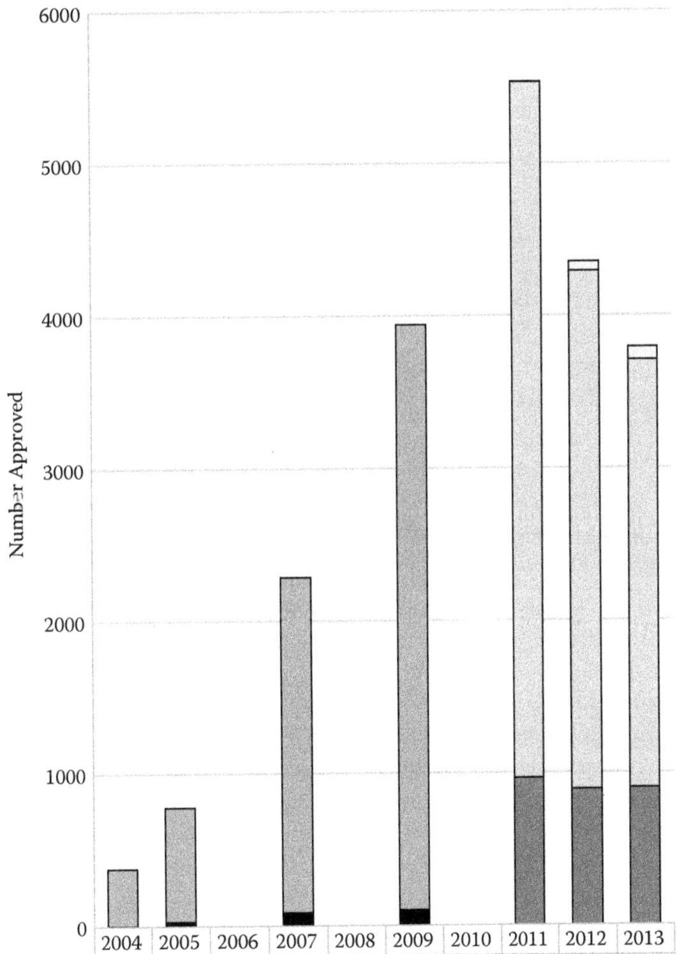

	2004	2005	2006	2007	2008	2009	2010	2011	2012	2013
□ O7 Typical (General) Notification	0	0	0	0	0	0	0	3	61	84
□ O7 Simple Notification	0	0	0	0	0	0	0	4567	3402	2807
■ O7 Scientific Research	0	0	0	0	0	0	0	964	890	899
□ O17 Exemption	375	750	0	2200	0	3840	0	0	0	0
■ O17 Normal Notification	2	29	0	86	0	100	0	0	0	0

FIGURE 3.5
Notifications and exemptions (China).

Order Number 7 reflects a tiered system for notifying the government of the manufacture or importation of new chemicals:

- *Simplified notifications*, for substances produced or imported at less than one tonne per year. Minimal data are required. These data include melting point, water solubility, and octanol–water partition coefficient. For organic substances, the simplified notification must also include data from a test of ready biodegradability and, if the substance is not readily biodegraded, a test report on acute aquatic toxicity (ideally to fish). An earthworm acute toxicity test may be required. For inorganic substances, testing on acute aquatic toxicity (and, under certain circumstances, acute toxicity to earthworms) is required [145,146].
- *Scientific research record notifications*, for substances intended solely for research purposes at less than one tonne per year. Minimal data are required.
- *Typical or general notifications*, which include series notifications of several substances with similar uses, joint notifications by more than one party, joint series notifications, repeat notifications, and renotification (when a substance moves to a higher tonnage band or uses change).

The amount of data required with each notification and the requirement to perform a risk assessment depends on the tonnage band. Table 3.8 summarizes the data requirements by annual production amount [145]. One can use existing data to exempt a filing from some of these data requirements under certain circumstances. The regulations do not allow for the use of QSAR [146].

The new Provisions require a risk assessment of new chemical substances to be manufactured at greater than 1 tonne per year, with chemicals management scaled to the level of concern [143]. The risk assessment report must include a hazard assessment, exposure assessment, and risk assessment. It must also specify risk controls.

The physicochemical hazard assessment must consider the potential threats from combustion, explosiveness, oxidizability, and metal corrosion. The human health risk assessment must qualitatively rank the hazard presented by the new substance, and the submitter must provide "the doses (concentration) of the new chemical substance … to match its no-hazard or minimum hazard status," akin to a DNEL as described in Chapter 2. The environmental hazard assessment must qualify the hazard level as severe, medium, or mild and a concentration akin to the PNEC, as discussed in Chapter 2, derived. The exposure assessment comprises two parts: a simple exposure forecast and assessment and a detailed exposure forecast and assessment. The latter includes a material balance on manufacturing or processing, and exposure scenarios for different exposure routes that describe the extent, range, and frequency of exposures. The results of the hazard

TABLE 3.8
Minimum Data Requirements for Typical Notification Under Order No. 7

Required Data	Grade 1 1–10 t/a	Grade 2 10–100 t/a	Grade 3 100–1,000 t/a	Grade 4 >1,000 t/a
Physicochemical Properties				
State of Substance	✓	✓	✓	✓
Melting/Freezing Point	S	S	S	S
Boiling Point	L	L	L	L
Density	L, S	L, S	L, S	L, S
Vapor Pressure	L	L	L	L
Surface Tension (If Applicable)	L	L	L	L
Water Solubility	L, S	L, S	L, S	L, S
Partition Coefficient n-Octanol/Water	L, S	L, S	L, S	L, S
Flash Point	G, L	G, L	G, L	G, L
Flammability	G, L, S	G, L, S	G, L, S	G, L, S
Explosive Limits	L, S	L, S	L, S	L, S
Auto-Ignition Temperature	G, L, S	G, L, S	G, L, S	G, L, S
Oxidizing Properties	G, L	G, L	G, L	G, L
Particle Diameter	S	S	S	S
pH	L	L	L	L
Toxicological Information				
Skin Irritation	✓	✓	✓	✓
Eye Irritation	✓	✓	✓	✓
Skin Sensitization	✓	✓	✓	✓
Mutagenicity	✓	✓	✓	✓
Acute Oral Toxicity	✓	✓	✓	✓
Acute Dermal Toxicity	✓	✓	✓	✓
Acute Inhalation Toxicity	✓	✓	✓	✓
Repeat Dose Oral Toxicity (28-Day)	✓	✓	✓	✓
90-Day Repeat Dose		✓ If warranted	✓	✓
Reproductive/Developmental Screening		✓		
Reproductive/ 2-Generation Reproductive Testing		✓ If warranted	✓	✓
Developmental			✓	✓
Toxicokinetics		✓ (Absorption)	✓	✓
Chronic Toxicity				✓
Carcinogenicity				✓

Continued

TABLE 3.8 (*Continued*)

Minimum Data Requirements for Typical Notification Under Order No. 7

Required Data	Grade 1 1–10 t/a	Grade 2 10–100 t/a	Grade 3 100–1,000 t/a	Grade 4 >1,000 t/a
Environmental Fate and Ecotoxicological Information				
Short-Term Invertebrate (*Daphnia*) Toxicity	✓	✓	✓	✓
Algal Growth Inhibition	✓	✓	✓	✓
Acute Fish Toxicity	✓	✓	✓	✓
Activated Sludge Respiration Inhibition	✓	✓	✓	✓
Daphnia Magna Reproduction Test		✓	✓	✓
14-Day Toxicity Testing on Fish		✓		
Chronic Toxicity to Fish			✓	
Bioaccumulation		✓	✓	✓
Degradation (Biotic)	✓	✓	✓	✓
Adsorption/Desorption	✓	✓	✓	✓
Acute Toxicity to Earthworms	✓	✓	✓	✓
Seed Germination/Root Elongation Test			✓	✓

assessments and exposure forecasts must be compared and "the risks of the new chemical substance to human health and the environment shall be described to reach a conclusion whether the risks are acceptable" considering the uncertainties. Finally, the registration submittal must describe risk control measures appropriate to occupational exposure, consumers' exposure, and environmental exposure [145].

An expert panel at MEP reviews the report and comments on suitable risk control measures. The expert panel also classifies new chemical substances as general new chemical substances or hazardous new chemical substances. The criteria for these classifications are not readily apparent, although hazardous substances reportedly include PBT substances [143]. Once a new substance is registered, the amended Provisions require that local environmental protection authorities supervise and inspect the manufacture, processing, and use of the chemical substances and report any problems to a higher authority [143,146].

Figure 3.5 shows the number and distribution of notifications in the first few years of the program established under Order Number 7 [147]. Of the 148 notifications approved for substances >1 tonne per year between 2011 and 2013, 18% were reportedly classified as general substances, 41% as hazardous substances, and 41% as priority hazardous substances [147]. These substances will be added to the IECSC five years from the date of the first production or import of the substance by the holder of the registration certificate [147].

In summary, the manufacture of new chemicals in China has only been regulated since 2003. Approximately 45,000 substances were listed as existing chemicals before the 2003 Provisions began chemical controls. The MEP accepted 217 notifications for new chemicals under Order Number 17 between 2004 and 2010, and 148 Typical Notifications under Order Number 7 from 2011 to 2013.

Some 3,000 of the approximately 47,000 existing chemical substances are classified as hazardous chemicals, but the level of characterization and control of the remainder is unclear. It is too early to determine the effects of the chemical control regulations promulgated between 2010 and 2013, or of the *Twelfth Five Year Plan Concerning the Prevention and Control of Chemical Environmental Risks.*

3.3.3 Recapitulation

- Chemical control regulations in China distinguish between substances in commerce when relevant regulations came into effect and chemicals newly manufactured or imported after that time.
- Approximately 45,000 chemicals were listed on the IECSC when the Provisions on the Environmental Administration of New Chemical Substances (SEPA Order Number 17) came into effect in 2003. China banned or restricted some of those existing chemicals based on international conventions and regulated the import/export and use of certain hazardous chemicals.
- China lists 162 chemicals, as of this writing, as banned or restricted. These substances can still be imported with a special permit.
- The Catalog of Hazardous Chemicals contains approximately 3,000 chemicals. Importers and manufacturers must register their hazardous chemicals, and manufacturers and vendors have to obtain licenses for manufacturing and sales, respectively. Hazardous chemicals must also be appropriately classified and labeled. Finally, companies that produce hazardous chemicals of high environmental concern must conduct an environmental risk assessment and report information on emissions to the MEP. Although these provisions impose some hazard communication requirements for hazardous chemicals, they do not seem to restrict or control their use per se.
- New chemical substances must be registered under the Provisions on Environmental Administration of New Chemical Substances. As initially promulgated in 2003, the Provisions required the submittal of data but not risk assessment. For newly registered substances, the regulation only required reporting and did not support the use of risk control measures or enforce any restrictions. As amended in 2010, the Provisions require an extensive data submittal and risk

assessment of new substances in commerce at greater than 1 tonne per annum. Risk control measures and reporting requirements can be imposed as a result of registration.

- The MEP accepted 217 Normal Notifications for new chemicals under the Provisions on Environmental Administration of New Chemical Substances between 2004 and 2010, and 148 Typical Notifications under the Provisions from 2011 to 2013; in addition, nearly 11,000 Simple Notifications have been accepted between 2011 and 2013.
- As this summary illustrates, chemical control regulations in China have evolved rapidly since 2000. Further evolution may occur as a result of the Twelfth Five Year Plan Concerning the Prevention and Control of Chemical Environmental Risks.

3.4 Voluntary Initiatives

In parallel with the laws and regulations described above, trade organizations and working groups have developed voluntary initiatives for product stewardship. Two of the best known are Responsible Care® and the Nano Risk Framework.

3.4.1 Responsible Care

In the early 1980s, the chemical industry in Canada recognized that it had a public relations problem: The public did not trust the chemical industry. The Canadian Chemical Producers' Association, now known as the Chemistry Industry Association of Canada, responded by creating the Responsible Care program, which developed voluntary guidelines for the safe and environmentally sound management of chemicals [148]. This started a global movement. By 2012, 55 chemical associations and more than 150 of the largest chemical companies participated in Responsible Care [149]. National associations coordinate Responsible Care; implementation in the United States, European Union, and China is described below.

3.4.1.1 United States

In the United States, the American Chemistry Council (then known as the Chemical Manufacturers' Association) adopted Responsible Care in 1988. Member companies adhered to 10 voluntary guiding principles of Responsible Care [150]. The original principles (which were updated in 1999 and 2008) were as follows:

1. Recognize and respond to community concerns about chemicals and our operations.
2. Develop and produce chemicals that can be manufactured, transported, used, and disposed of safely.
3. Make health, safety, and environmental considerations a priority in our planning for all existing and new products and processes.
4. Report promptly to officials, employees, customers, and the public, information on chemical-related health or environmental hazards and recommend protective measures.
5. Counsel customers on the safe use, transportation, and disposal of chemical products.
6. Operate our plants and facilities in a manner that protects the environment and the health and safety of our employees and the public.
7. Extend knowledge by conducting or supporting research on the health, safety, and environmental effects of our products, processes, and waste materials.
8. Work with others to resolve problems created by past handling and disposal of hazardous substances.
9. Participate with government and others in creating responsible laws, regulations, and standards to safeguard the community, workplace, and environment.
10. Promote the principles and practices of Responsible Care by sharing experiences and offering assistance to others who produce, handle, use, transport, or dispose of chemicals.

(Emphasis added to the two principles most relevant to the analyses in this book.*)

To support these principles, companies adopted the Responsible Care Codes of Management Practices. The Product Stewardship Code created in 1992 [152] was intended to make "make health, safety and environmental protection an integral part of designing, manufacturing, marketing, distributing, using, recycling and disposing of our products." Participating companies would have an ongoing product stewardship process that reflected management commitment to product stewardship. The product stewardship process would include the collection and maintanence of information on hazards and exposures, risk assessment, and risk management measures. Participating companies would establish goals for implementing product stewardship and measure performance against these goals.

* ACC has modified the principles to read as follows: "To design and develop products that can be manufactured, transported, used and disposed of or recycled safely" [151].

Initially, companies simply signed on to Responsible Care as a condition of membership in the CMA/ACC. The program evolved to become more rigorous. In 1993, the ACC adopted performance metric reporting, instituted deadlines for implementing the codes, and initiated a voluntary verification process. Four years later, ACC transitioned from Codes to a Responsible Care Management System and formalized third-party certification; in 2003, Responsible Care companies began public reporting of peformance metrics [153]. In 2012, ACC formalized the Product Safety Code, the successor to the Product Stewardship Code. With a phase-in period of four years, the Product Safety Code outlines a series of management practices that require member companies to [154]:

- Manage the safety of their products throughout the life cycle
- Undertake scientific analyses of their products, with close consideration of how they are used by consumers, especially children
- Provide public access to product safety information
- Share information along the supply chain so that products can be safely used
- Continuously evaluate new information that may have product safety implications
- Take corrective measures if they discover improper practices involving a product

How well has the Responsible Care program worked in the United States? OECD has noted the challenge of measuring the effectiveness of the program, remarking that "Responsible Care's focus on management systems makes progress more difficult to measure than initiatives that contain quantitative performance goals " [155, p. 4]. With that caution in mind, let's begin by looking at reported data.

Responsible Care companies report multiple performance metrics, as indicated in Table 3.9 [156]. Two of the metrics listed in Table 3.9 qualitatively address product stewardship concerns.* When asked, "Does your company have a process to prioritize its chemicals?" 74% of responding companies answered affirmatively in 2007 and 90% in 2012. In response to the question, "Is a summary of the process used to prioritize your chemicals available to the public?" 35% of responding companies answered "Yes" in 2007 and 50% in 2012 [158]. The response rate is not clear from the information provided, nor is information presented on the implementation or effectiveness of the prioritzation processes.

Four of the quantitative metrics in Table 3.9—hazardous air pollutants (HAP) emissions, process safety incidents, distribution incidents, and OSHA-reportable incidents—speak to the two product stewardship goals in

* With respect to the product stewardship goal of amassing and sharing hazard information, the International Council of Chemical Associations (ICCA) has also noted the contribution of ACC companies, which are all participants in Responsible Care, to the HPV data-gathering effort in the United States [157].

TABLE 3.9

Responsible Care Metrics

Category	Metric Tracked by ACC	Metric Tracked by Cefic*
Environment—Air	Releases of hazardous air pollutants (HAP) Emissions of sulfur oxides (SOx) Emissions of nitrous oxides (NOx)	Emissions of nonmethane volatile organic compounds (NMVOCs) Sulfur dioxide emissions Nitrous oxide emissions
Environment— Water	Net water consumption	Releases of nitrogen Releases of phosphorus Chemical oxygen demand
Environment— Waste	—	Hazardous waste Nonhazardous waste
Energy	Greenhouse gas emissions Energy efficiency	Greenhouse gas emissions
Safety	Number of process safety incidents Distribution incidents reportable to the Department of Transportation Lost workday incidence and workplace fatalities reportable to the Occupational Safety and Health Administration (OSHA) Product safety: • Process in place to prioritize chemicals according to potential risk • Summary of process available to public	Fatalities Lost-time injury frequency
Accountability	Community outreach Emergency response initiatives	—

* Cefic: European Chemical Industry Council.

Principles 2 and 3, above. (Two, SOx and NOx emissions, are arguably relevant.) All reference an existing regulatory program such as the Clean Air Act or Occupational Safety and Health Act rather than an industry initiative under Responsible Care. None pertains directly to the life cycle thinking about chemicals and the products that contain them, which was embodied in the original principles.

The quantitative metric most relevant to the questions explored in this book is the release of HAPs, which are chemicals that the US EPA has identified as substances known or suspected to cause cancer or other serious health effects or adverse environmental effects. Releases of HAPs must be quantified under the US EPA's Toxics Release Inventory (TRI). ACC has reported that between 1988 and 2011, Responsible Care companies reduced HAP emissions by 76% [159].

Studies have examined the effectiveness of Responsible Care in the United States by evaluating publicly available TRI data. King and Lenox [160] opined that industry efforts at self-regulation such as Responsible Care can be ineffective because they do not contain explicit sanctions for poor performance; in fact, antitrust regulations prevent trade associations from imposing certain sanctions. This research team aggregated TRI data from 1987–1996 to examine the performance of 1,500 chemical companies at both the facility level and corporate level, on the basis that TRI data served as a proxy for measuring environmental performance relative to releases of toxic chemicals during manufacture and distribution. Of the companies evaluated, 150 participated in Responsible Care in 1990 and 160 by 1996. Industry as a whole reduced the total toxicity-weighted emissions by nearly 50% during the time period studied. However, the researchers found that participation in Responsible Care did not positively influence the rate of improvement in TRI emissions.

In a subsequent study, Gamber-Rabindran and Finger reviewed TRI data between 1988 and 2001 for 1,759 firms which owned a total of 3,278 facilities. The study concluded that plants participating in Responsible Care did not reduce their pollution relative to statistically equivalent nonparticipants. The authors noted that they examined data from the period before ACC instituted a verification program, which may have affected subsequent performance [161].

The fundamental principles of Responsible Care and instruments such as the Product Stewardship Code and the Product Safety Code clearly support product stewardship goals and life cycle thinking about product design. However, as described above, the accessible aggregated data about efforts in the United States do not quantitatively demonstrate the program's effectiveness.

3.4.1.2 European Union

Trade associations for countries in Europe have been signing on to Responsible Care since 1986 [149]. In Europe, the European Chemical Industry Council (Cefic) promotes Responsible Care, which is actually implemented by national member federations. Cefic collects data from 22 countries on 16 key performance indicators (KPIs). Eleven of the KPIs are reported on Cefic's website, as shown in Table 3.9 [162]. One of the indicators most relevant to the questions explored in this book is the release of volatile organic compounds (VOCs) to the air during manufacturing. More specifically, Responsible Care companies in Europe report the discharges of nonmethane VOCs. These compounds reflect the potential for photochemical ozone creation that may cause respiratory problems and ecological damage. Cefic reports that VOC discharges decreased by approximately 34 wt% between 2003 and 2008 [163]. It is not clear how much of this decrease can be attributed to participation in Responsible Care.

3.4.1.3 China

Responsible Care programs have been active in the West for over 25 years, however, the China Petroleum and Chemical Industry Federation (CPCIF;

formerly the China Petroleum and Chemical Industry Association or CPCIA) began to explore Responsible Care in China in the mid-2000s and became a member (observer) of the International Council of Chemical Associations (ICCA) in 2010. Activities to date have included development of the *Guidelines for Implementation of Responsible Care* (industrial standard) and the engagement and training of chemical companies. The 2011 Guidelines, which reinforce the importance of relevant government regulations and standards, include six codes that pertain to product stewardship (including risk characterization, risk management, and communication), process safety, occupational health and safety, pollution prevention and control, storage and transport safety, community awareness, and emergency response [164–167].

In addition to the efforts by the CPCIF, the Association of International Chemical Manufacturers (AICM) participates in Responsible Care in China. These international companies produce 10–15% of the chemical products manufactured in China. Participating AICM members implement Responsible Care principles at their own operations and support Chinese companies introducing Responsible Care [168,169]. Support and expansion of Responsible Care in China is a stated goal for ICCA [170].

3.4.1.4 Looking Forward

The Responsible Care program will likely continue to evolve. ICCA supports the goal adopted by the World Summit on Sustainable Development in 2002: "that by 2020 chemicals are produced and used in ways that minimize significant adverse impacts on human health and the environment" [170]. How Responsible Care will contribute to that goal, and how progress will be measured and reported, remain to be seen.

3.4.2 Nano Risk Framework*

The Environmental Defense Fund, a US-based nonprofit environmental advocacy group, and the multinational chemical company DuPont collaborated to develop the Nano Risk Framework [171]. In the words of the developers,

> The purpose of this Framework is to define a systematic and disciplined process for identifying, managing, and reducing potential environmental, health, and safety risks of engineered nanomaterials across all stages of a product's "lifecycle"— its full life from initial sourcing through manufacture, use, disposal or recycling, and ultimate fate. The Framework offers guidance on the key questions an organization should consider in developing applications of nanomaterials, and on the information needed to make sound risk evaluations and risk-management decisions. The Framework allows users flexibility in making such decisions in the

* Note: This section is adapted from Section 11.2.2 of *Nanotechnology and the Environment* (Taylor & Francis/CRC Press, 2009) and is used with permission.

> presence of knowledge gaps—through the application of reasonable assumptions and appropriate risk-management practices. Further, the Framework describes a system for guiding information generation and updating assumptions, decisions, and practices with new information as it becomes available. And the Framework offers guidance on how to communicate information and decisions to key stakeholders.

The Framework differs from a complete life cycle analysis in that it focuses on potential environmental, health, and safety risks. It does not consider resource inputs. The Nano Risk Framework comprises six steps, as described briefly below.

1. *Describe Material and Application.* This step generates an overview of the physical and chemical properties of the material, sources, and manufacturing processes, and possible uses. The overview includes existing materials that the nanomaterial may replace, and bulk counterparts of the nanomaterial.
2. *Profile Life Cycle(s).* This step includes three components. Each relies on compiled "base set" data to define the characteristics and hazards of a nanomaterial. Where those data are not available, the Framework suggests using reasonable worst-case default values or assumptions. Analysts can replace those default values with actual data as they become available. This approach will provide an initially conservative estimate of risk that can be refined if appropriate.
   a. *Profile Life Cycle Properties.* Develop base set data on physical and chemical properties of the nanomaterial, including property changes throughout the full product life cycle.
   b. *Profile Life Cycle Hazards.* Characterize the potential hazards to human health, the environment, and safety from exposure to this material throughout its life cycle. In this step, analysts compile four base sets of data: health hazards, environmental hazards, environmental fate, and safety. Standard methods are not yet available to measure some of these base set parameters for nanomaterials. Base set data on health hazards include short-term toxicity, skin sensitization/irritation, skin penetration, genetic toxicity tests, and other data. Base set environmental hazard data include acute aquatic toxicology and terrestrial toxicology (i.e., earthworms and plants), and may include additional data if needed. Recommended base set data on the environmental fate of nanomaterials include physical–chemical properties, adsorption–desorption coefficients (soil or sludge), and nanomaterial aggregation or disaggregation in applicable exposure media. They also include data pertaining to persistence, characterizing

biodegradability, photodegradability, hydrolysis, and bioaccumulation. Finally, base set safety hazard data include flammability, explosivity, incompatibility, reactivity, and corrosivity.

c. *Profile Life Cycle Exposure.* Quantify the potential for human and environmental exposures throughout the product life cycle. This definition is deceptively simple. The analyst must consider opportunities for direct contact or release to the environment at multiple stages: manufacture, processing, use, distribution/storage, and postuse disposal, reuse, or recycling.

3. *Evaluate Risks.* The information collected in Steps 1 and 2 is combined to estimate the risks to human health and the environment for each life cycle stage. Depending on the availability of base set data, the initial estimates may range from qualitative to quantitative. The analyst must determine gaps in the life cycle profiles and either generate data to fill the gaps or make reasonable worst-case assumptions.
4. *Assess Risk Management.* For each life cycle stage, determine the actions needed to reduce and control risks from known and reasonably anticipated activities. These actions could include product modifications, engineering or management controls, protective equipment, or risk communication such as warning labels. The product developer might even decide to abandon the product.
5. *Decide, Document, and Act.* At this stage, a review team critically analyzes the results to decide how to proceed. The team documents and communicates the results, and determines the course of action for refining or updating the conclusions.
6. *Review and Adapt.* This step ensures that the risk characterization and risk management protocols continue to evolve as new information becomes available.

DuPont applied this framework to DuPont™ Light Stabilizer 210 [172] in 2007, describing the material and application as follows. This product is a surface-treated form of titanium dioxide that absorbs and scatters ultraviolet (UV) light. Addition of this product to a polymer (at $\leq 3\%$ of the end product) protects the material from UV damage when exposed to sunlight. Potential applications include outdoor furniture, toys, and sheeting to protect greenhouses.

The life cycle profile of this product, reflecting the data available at the time, noted the following. DuPont Light Stabilizer 210 is a white powder with particle sizes centered in the range of 130–140 nm. 10–20 wt% falls within the nano range, that is, <100 nm. The particles are dense polyhedral TiO_2 crystals surface treated to control chemical reactions. The particles cannot be broken down by mechanical action and their composition will not substantially change throughout the life cycle.

Titanium occurs naturally in the environment. No established analytical method can distinguish between the titanium in DuPont Light Stabilizer 210 and naturally occurring titanium.

Toxicity studies showed no significant difference between the effects of DuPont Light Stabilizer 210 and pigmentary TiO_2. Aquatic toxicity testing demonstrated low hazard to fish and invertebrates and indicated medium concern for algae, potentially due to the light-blocking effects. No accepted protocols for assessing the bioaccumulation potential of nanomaterials exist.

Worker exposure should be low under normal operating conditions. Monitoring during production and handling indicated that airborne concentrations were below the acceptable exposure limit of 2 mg/m^3. If exposure limits were exceeded, workers were to don half-mask respirators with P100 filters according to the safety precautions specified for DuPont Light Stabilizer 210.

Exposure was expected to be low throughout the product life cycle for several reasons: potential worker exposure is well managed; due to the low production, use of engineering controls, and properties of the material, releases to the environment should be minimal; and the polymer end product should retain the DuPont Light Stabilizer 210 unless incinerated. Emissions from incineration should be low due to the low concentrations and emission controls on incinerators.

The evaluation of risks reflected the finding that toxicity studies showed no significant difference between the effects of exposure to DuPont Light Stabilizer 210 and pigmentary TiO_2; each showed low hazard. Further, exposure should be limited. Therefore, "there are no substantive risk issues associated with manufacture, processing, use or disposal of DuPont™ Light Stabilizer."

The assessment of risk-management measures recommended few additional risk management measures. Those included personnel scheduling and monitoring during nonroutine activities, and developing recycling procedures. Some additional toxicity testing was contemplated.

Regarding the fifth step, to decide, document, and act, the review team accepted the recommendations of risk management measures and approved moving forward to product announcement and commercialization. DuPont scheduled reviews of DuPont Light Stabilizer 210 in 2009 and then every four years thereafter, and noted that "as needed" risk reviews would occur if triggered by a change in applications, new information on hazard, or higher-than-anticipated production.

The results of subsequent reviews do not appear to be readily available, although a publication by a DuPont scientist in 2013 reaffirmed the essential toxicological and ecotoxicological findings [173]. A search of DuPont's product website in April 2014 did not identify a current listing for Light Stabilizer 210.

3.5 Crossing Borders: Influence of Global Trade

National regulations govern the manufacture and import of chemical substances in a given country, and some companies choose to adhere to voluntary codes of conduct developed by trade associations typically based in their geographic region. A third factor has begun to influence product stewardship: the global marketplace. Consider the following examples.

A product made within a country with no chemical control regulations is shipped to the United States for sale. Under Section 13 of TSCA, the importer must certify that the chemical substances in the shipment comply with all applicable rules under TSCA or else that the chemicals are not subject to TSCA. Other geographies have similar restrictions on imports. Although not foolproof, and reliant on the importer's accurate knowledge of the product formulation, this requirement can affect the use of certain chemicals in a country with no applicable regulations of its own.

"Retail regulation," based on customer deselection of products containing certain chemicals, can affect the use of those chemicals. Since 2001, SC Johnson has used the Greenlist™ process to classify substances considered for use in its products according to their impact on human health and the environment, and used that process as a basis for product design [174]; the Greenlist process has become part of its brand. Whole Foods banned bisphenol A from baby bottles and children's cups on its store shelves in 2006, and other retailers followed suit. In 2011, Walmart decided to stop selling products containing the flame retardant known as polybrominated diphenyl ethers, or PBDEs, and has since moved to eliminate other substances from its shelves [175]. This list is by no means complete, but illustrates how choices based on consumer preferences can affect the use of chemicals that are not banned or restricted by regulation. Such initiatives can control the use of chemicals regardless of promulgated regulations.

3.6 Synthesis

This book began with the deliberately naïve question of why, given our technical knowledge and the growth of environmental regulations, anthropogenic chemicals can be found throughout the environment. This chapter explored chemical control regulations in three of the major chemical-producing economies as well as voluntary efforts and market-driven choices regarding chemical use. The answer to the question, "Why?" reflects when regulations came into effect, how they were developed and implemented, and the balance between risks and benefits.

Governments moved to regulate chemical manufacture and use relatively late in the Industrial Revolution. Initial legislation in the United States, European Union, and China reflected the pragmatic decision to distinguish between chemicals in commerce at the time from chemicals newly brought to market afterwards. In consequence, tens of thousands of chemicals already on the market in each geography were "grandfathered" with no immediate requirement for assessment or control other than that afforded by existing end-of-pipe pollution control regulations.

Assessment and control of new chemical substances required regulators to develop new scientific paradigms. They built tools to characterize the fate and transport of chemicals in the environment and to understand the toxicity and ecotoxicity of chemical substances so that their use could be appropriately controlled if necessary. In hindsight, early efforts were often crude and lacked the sensitivity to characterize some of the subtle consequences of chemical exposure. But the protocols now taken for granted had to be developed and tested from first principles, often with a paucity of data. Even after 40 years of refinement, regulatory paradigms cannot address all of the complexities of global chemical usage. Consider, for example, that chemical notification protocols in the United States and the European Union reflect exposure characterization by an individual manufacturer or importer for a single chemical substance; the legislative and technical tools available to regulators do not allow them to assess the results of cumulative exposure and reflect those results in chemical controls. Further, regulations promulgated 10 to 40 years ago must adapt to changes in material science and engineering over time, and such changes in the marketplace can outpace the changes in regulations. The US EPA began to grapple with the regulation of genetically engineered substances under TSCA circa 1983 [176] and published new regulations under TSCA applicable to biotechnology submissions in 1997 [177]. More recently, as manufacturers have exploited the unique properties of nanomaterials to bring hundreds of nanotechnology-enabled products to the marketplace, regulators around the world have struggled to adapt chemical registration programs to the unique properties of these materials.

Legislators and regulators have also borne responsibility for the economic health of their countries, and reflected in chemical control legislation the need to balance the effects of chemical manufacture and use on human health and the environment against the economic effects of limiting chemical manufacture and use. That balance reflected judgments, not always transparent, about the acceptable level of risk to human health and the environment. And the decision whether to control the manufacture and use of a chemical substance could be tipped by the opportunity to control exposures by other national laws limiting air or water pollution, or requiring treatment of drinking water. Legislative mandates to keep this balance have sometimes resulted in deliberations that took decades before controls were imposed.

In short, environmental conditions reflect a series of decisions, sometimes imperfectly made and based on limited information. The next chapter provides a series of extended case studies to illustrate the consequences in more detail.

References

[1] United Nations Environment Program. Undated. *Rotterdam Convention*. Available at: http://www.pic.int/(July 27, 2013).
[2] United Nations Environment Programme (UNEP). 2009. *Stockholm Convention on Persistent Organic Pollutants (POPs) as Amended in 2009 Text and Annexes*. Available at: http://chm.pops.int/Convention/tabid/54/Default.aspx (accessed July 27, 2013).
[3] Massey, R. and Jacobs, M. 2013. *United Nations Global Chemicals Outlook - Towards Sound Management of Chemicals*. Chapter I: Trends and Indicators. Environment Programme. Available at: http://www.unep.org/hazardoussubstances/Portals/9/Mainstreaming/GCO/The%20Global%20Chemical%20Outlook_Full%20report_15Feb2013.pdf (accessed July 2013).
[4] American Chemistry Council. 2012. 2012 Guide to the Business of Chemistry. *Business of Chemistry (Annual Data 2012)*.
[5] Council on Environmental Quality (CEQ). 1971. Toxic Substances, reprinted in *Staff House Committee on Interstate and Foreign Commerce, 94th Congress, 2nd Session*, Legislative History of the Toxic Substances Control Act (TSCA Legislative History) at 760 (Comm. Print 1976), pp. 755–788.
[6] President's Advisory Council on Executive Organization. 1970. *Memorandum for the President. Subject: Federal Organization for Environmental Protection. ("Ash Council Memo")*. April 29, 1970. Available at: http://www2.epa.gov/aboutepa/ash-council-memo (accessed June 21, 2013).
[7] US EPA. 1997. *Chemistry Assistance Manual for Premanufacture Notification Submitters, EPA 744-R-97-003, March 1997. Appendix A*. Available at: http://www.epa.gov/oppt/newchems/pubs/chem-pmn/appendix.pdf (accessed June 21, 2013).
[8] Library of Congress. 1976. *Legislative History of the Toxic Substances Control Act, Together with a Section-by-Section Index*. Prepared by the Environment and Natural Resources Policy Division of the Library of Congress for the House Committee on Interstate and Foreign Commerce. December.
[9] US Bureau of Labor Statistics. *Labor Force Statistics from the Current Population Survey. Databases, Tables & Calculators by Subject*. Available at: http://data.bls.gov/pdq/SurveyOutputServlet (accessed June 27, 2013).
[10] BP. 2013. *BP Statistical Review of World Energy*. Available at: www.bp.com/en/global/corporate/about-bp/statistical-review-of-world-energy-2013/statistical-review-1951-2011.html (accessed June 27, 2013).
[11] US Department of State, Office of the Historian. Undated. *OPEC Oil Embargo, 1973-1974*. Available at: http://history.state.gov/milestones/1969-1976/OPEC (accessed June 28, 2013).

[12] Office of the White House Press Secretary. 1976. *Fact Sheet – S. 3149 – The Toxic Substances Control Act*. October 12, 1975. Digitized from Box 32 of the White House Press releases at the Gerald R. Ford Presidential Library. Available at: http://www.fordlibrarymuseum.gov/ford_full_search.html (accessed June 21, 2013).

[13] Toxic Substances Control Act. 15 U.S.C., Title I, Section 3. Definitions.

[14] US EPA. 2013. *What Is the TSCA Chemical Substance Inventory?* Web page - Last updated on April 03, 2013. Available at: http://www.epa.gov/oppt/newchems/pubs/invntory.htm (accessed June 21, 2013).

[15] US EPA. 2013. *Chemical Data Reporting. Fact Sheet: Basic Information*. EPA Publication 740K13001. June 2013. Available at: http://www.epa.gov/oppt/cdr/pubs/guidance/1st_cdr_basic_factsheet.pdf (accessed July 14, 2013).

[16] US EPA. 2004. *Status and Future Directions of the High Production Volume Challenge*. Available at: http://www.epa.gov/hpv/pubs/general/hpvreport.pdf (accessed July 12, 2013).

[17] National Pollution Prevention and Toxics Advisory Committee (NPPTAC). 2005. *Initial Thought-Starter: How Can EPA More Efficiently Identify Potential Risks and Facilitate Risk Reduction Decisions for Non-HPV Existing Chemicals*? October. Available at: http://www.epa.gov/oppt/npptac/pubs/finaldraftnonhpvpaper051006.pdf (accessed July 14, 2013).

[18] US General Accounting Office (GAO). 1989. *Report to the Congress of the United States by the Comptroller General: EPA Is Slow to Carry Out Its Responsibility to Control Harmful Chemicals*. Available at: http://www.gao.gov/assets/140/130786.pdf (accessed July 14, 2013).

[19] US GAO. 1980. *Study by the Staff of the U.S. General Accounting Office. Environmental Protection Issues in the 1980s*. Available at: http://www.gao.gov/assets/80/79279.pdf (accessed July 14, 2013).

[20] US EPA. 2011. *Data Development (Testing)*. Web page – last updated on April 27, 2011. Available at: http://www.epa.gov/opptintr/chemtest/pubs/data.html (accessed July 7, 2013).

[21] US EPA. 2004. *Status and Future Directions of the High Production Volume Challenge*. Available at: http://www.epa.gov/hpv/pubs/general/hpvreport.pdf (accessed July 12, 2013).

[22] US EPA. 2012. *High Production Volume (HPV) Challenge - Basic Information*. Web page last updated April 12. 2012. Available at: http://www.epa.gov/hpv/pubs/general/basicinfo.htm (accessed July 12, 2013).

[23] US EPA. 2013. *High Production Volume Information System (HPVIS) - HPV Chemical Hazard Characterizations*. Web page last updated July 12, 2013. Available at: http://iaspub.epa.gov/oppthpv/hpv_hc_characterization.get_report?doctype=2 (accessed July 12, 2013).

[24] US EPA. 2009. *Screening-Level Hazard Characterization, Mononitroanilines Category*. Available at: http://www.epa.gov/chemrtk/hpvis/hazchar/Category_Mononitroanilines_Sept2009.pdf (accessed July 12, 2013).

[25] US EPA. 2012. *Existing Chemicals Program: Strategy. February 2012*. Available at: http://www.epa.gov/oppt/existingchemicals/pubs/Existing_Chemicals_Strategy_Web.2-23-12.pdf (accessed July 12, 2013).

[26] US EPA. 2014. *TSCA Work Plan Chemicals: 2014 Update. October 2014*. Available at: http://www.epa.gov/oppt/existingchemicals/pubs/TSCA_Work_Plan_Chemicals_2014_Update-final.pdf (accessed November 2, 2014).

[27] US EPA. 2012. *Existing Chemicals Action Plans.* Web page last updated July 18, 2013. Available at: http://www.epa.gov/oppt/existingchemicals/pubs/ecactionpln.html (accessed July 12, 2013).

[28] US GAO. 1984. *EPA's Efforts to Identify and Control Harmful Chemicals in Use.* GAO/RCED-84-100. June 13, 1984. Available at: http://www.gao.gov/assets/150/141813.pdf (accessed July 12, 2013).

[29] 44 FR 202: 60061-60068. *Commercial and Industrial Use of Asbestos Fibers; Advance Notice of Proposed Rulemaking.* October 17, 1979]

[30] 54 FR 132: 29460 – 29513. 40 CFR Part 763. *Asbestos; Manufacture, Importation, Processing, and Distribution in Commerce Prohibitions*

[31] Stephenson, J. 2009. *Testimony before the Committee on Environment and Public Works, U.S. Senate. Observations on Improving the Toxic Substances Control Act.* Statement of John Stephenson, Director, Natural Resources and Environment. December 2, 2009. Available at: http://www.gao.gov/assets/130/123792.pdf. (accessed July 12, 2013).

[32] Hegstad, M. 2013. *In TSCA Debate, Industry Seeks To Limit Harmful Exposures, Not Hazards.* InsideEPA.com, March 4, 2013. Available at: http://insideepa.com/Inside-EPA-General/Public-Content-ACC/in-tsca-debate-industry-seeks-to-limit-harmful-exposures-not-hazards/menu-id-1026.html (accessed July 12, 2013).

[33] Roberts, J.A. 2010. *From Inception to Reform: An Oral History Project of the Toxic Substances Control Act.* Center for Contemporary History and Policy, Chemical Heritage Foundation. Available at: http://www.chemheritage.org/research/policy-center/projects/tsca-oral-history.aspx and http://www.elpnet.org/webinar-archive (accessed July 7, 2013).

[34] US GAO. 2005. *Options Exist to Improve EPA's Ability to Assess Health Risks and Manage Its Chemical Review Program.* GAO-05-458. Available at: http://www.gao.gov/products/GAO-05-458 (accessed July 14, 2013).

[35] US EPA. 2009. *Premanufacture Notice for New Chemical Substances.* EPA Form 7710-25 (Rev 6-09).

[36] US EPA. 1997. *Chemistry Assistance Manual for Premanufacture Notification Submitters.* Chapter 1. EPA 744-R-97-003. March, 1997. Available at: http://www.epa.gov/oppt/newchems/pubs/chem-pmn/chap1.pdf (accessed June 28, 2013.)

[37] Owens, S. 2013. Statement issued through press release issued by Frank Lautenberg, U.S. Senator for New Jersey: *Top EPA Toxics Officials Under Obama & Bush Administrations Hail Lautenberg-Vitter Legislation to Reform Nation's Chemical Laws.* May 23, 2013. Available at: http://www.lautenberg.senate.gov/newsroom/record.cfm?id=342882 (accessed June 28, 2013).

[38] US GAO. 2013. *Report to Congressional Requesters. TOXIC SUBSTANCES - EPA Has Increased Efforts to Assess and Control Chemicals but Could Strengthen Its Approach.* GAO-13-249. March, 2013. Available at http://www.gao.gov/assets/660/653276.pdf (accessed June 28, 2013).

[39] US EPA. 2012. *Chemical Categories Report.* Web page last updated April 3, 2013. Available at: http://www.epa.gov/oppt/newchems/pubs/chemcat.htm (accessed June 28, 2013).

[40] US EPA. 2012. *Estimation Program Interface (EPI) Suite Version 4.11* (November, 2012). Available at: http://www.epa.gov/opptintr/exposure/pubs/episuite.htm (accessed June 28, 2013).

[41] US GAO. 2001. *Chemical Risk Assessment: Selected Federal Agencies' Procedures, Assumptions, and Policies.* Appendix II - Chemical Risk Assessment at the Environmental Protection Agency. GAO-01-810. Available at: http://www.gao.gov/new.items/d01810.pdf (accessed June 29, 2013).

[42] US EPA. 2012. *Sustainable Futures/P2 Framework Manual 2012* EPA-748-B12-001. Chapter 2. http://www.epa.gov/oppt/sf/pubs/sf-p2-fr-methods.pdf (accessed June 29, 2013).

[43] US EPA. 1997. *1997 Chemical Engineering Branch Compilation of Generic Scenarios (for Industry-Specific Workplace Release and Exposure Estimation).* Available at: http://www.epa.gov/oppt/exposure/pubs/guidance.htm (accessed July 2013).

[44] US EPA. 1991? Surfactants-Household Cleaners.pdf. File included in the *1997 Chemical Engineering Branch Compilation of Generic Scenarios [for Industry-Specific Workplace Release and Exposure Estimation].* Available at: http://www.epa.gov/oppt/exposure/pubs/guidance.htm (accessed June 28, 2013).

[45] US EPA. 2012. *Exposure Assessment Tools and Models.* Web page last updated September 28, 2012. Available at: http://www.epa.gov/opptintr/exposure/ (accessed June 29, 2013).

[46] US EPA. 2004. *ChemSTEER Overview - Chemical Screening Tool for Exposures and Environmental Releases.* Office of Pollution Prevention & Toxics, Chemical Engineering Branch, Washington, DC. May 2004. Beta Version. Available at: http://www.epa.gov/oppt/exposure/pubs/chemsteerslideshow.htm (accessed June 28, 2013).

[47] US EPA. 2010. *Chemical Screening Tool for Exposures & Environmental Releases.* Web page last updated May 12, 2010. Available at: http://www.epa.gov/oppt/exposure/pubs/chemsteer.htm (accessed June 28, 2013).

[48] Nabholz, J.V., Miller, P., and Zeeman, M. 1993. *Environmental Risk Assessment of New Chemicals Under the Toxic Substances Control Act (TSCA) Section Five. Environmental Toxicology and Risk Assessment, ASTM STP* 1179, W. G. Landis, J. S. Hughes, and M. A. Lewis (Eds.). Philadelphia: American Society for Testing and Materials, pp. 40–55.

[49] US EPA. 2010. *Exposure and Fate Assessment Screening Tool Version 2.0* (E-FAST V2.0). Web page last updated May 12, 2010. Available at: http://www.epa.gov/oppt/exposure/pubs/efast.htm (accessed June 28, 2013).

[50] US EPA. 1997. *Chemistry Assistance Manual for Premanufacture Notification Submitters* (U.S. EPA 744-R-97-003, March 1997). Available at: http://www.epa.gov/opptintr/newchems/pubs/chem-pmn/index.htm (accessed June 28, 2013).

[51] US EPA. 2013. *TSCA 5(e) Exposure-Based Policy: Criteria.* Last updated April 3, 2013. Available at: http://www.epa.gov/oppt/newchems/pubs/expbased.htm (accessed June 28, 2013).

[52] Elkins, C. 1988. *Chemical Manufacturers Association Letter*. From Charles L. Elkins, Director, U.S. EPA Office of Toxic Substances to Geraldine V. Cox, Ph.D., Vice President - Technical Director, Chemical Manufacturers Association. September 22, 1988. Available at: http://www.epa.gov/oppt/newchems/pubs/cmaxpltr.htm (accessed June 24, 2013).

[53] US EPA. 2012. *TSCA 5(e) Exposure-Based Policy: Testing*. Web page updated on September 28, 2012. Available at: http://www.epa.gov/oppt/newchems/pubs/expbasedtesting.htm (accessed January 22, 2013).

[54] US EPA. 2010. *TSCA New Chemicals Program (NCP) Chemical Categories* (August 2010). Available at: http://www.epa.gov/oppt/newchems/pubs/npcchemicalcategories.pdf (accessed June 29, 2013).

[55] Mayo-Bean, K. and Morana, K. 2012. *AIM Standalone Public Version Users' Manual.* User's manual can be downloaded with the AIM software from http://www.epa.gov/oppt/sf/tools/aim.htm (accessed June 29, 2013).

[56] US EPA. 1999. *Category for Persistent, Bioaccumulative, and Toxic New Chemical Substances. Federal Register: November 4, 1999* (Volume 64, Number 213).

[57] US EPA. 2012. *PBT Profiler*. http://www.epa.gov/oppt/sf/tools/pbtprofiler.htm (accessed June 29, 2013).

[58] US EPA. 2012. *OncoLogic™ - A Computer System to Evaluate the Carcinogenic Potential of Chemicals*. Web page updated on May 29, 2012. Available at: http://www.epa.gov/oppt/sf/pubs/oncologic.htm (accessed June 29, 2013).

[59] US EPA. 2012. *Using NonCancer Screening Within the SF Initiative*. Web page last updated May 29, 2012. Available at: http://www.epa.gov/opptintr/sf/pubs/noncan-screen.htm (accessed June 29, 2013).

[60] US EPA. 2012. *Estimation Program Interface (EPI) Suite Version 4.11 (November, 2012). Users' Manual*. Available at: http://www.epa.gov/opptintr/exposure/pubs/episuite.htm (accessed June 28, 2013).

[61] US GAO. 2001. *Report to Congressional Requesters: Chemical Risk Assessment - Selected Federal Agencies' Procedures, Assumptions, and Policies. Appendix II Chemical Risk Assessment at the Environmental Protection Agency*. GAO-01-810. Available at: http://www.gao.gov/products/GAO-01-810 (accessed July 12, 2013).

[62] US EPA. 2013. *Making a Finding on Unreasonableness of Risk*. Web page last updated April 3, 2013. Available at: http://www.epa.gov/oppt/newchems/pubs/unrerisk.htm (accessed June 29, 2013).

[63] US EPA. 2011. *Exposure Factors Handbook*. Available at: http://www.epa.gov/ncea/efh/report.html (accessed December 6, 2013).

[64] US EPA. 2013. *Drinking Water Contaminants*. Web page last updated June 3, 2013. Available at: http://water.epa.gov/drink/contaminants/#List (accessed December 6, 2013).

[65] Todhunter J.A. 1983. Risk management strategy under the Toxic Substances Control Act and the Federal Insecticide, Fungicide, and Rodenticide Act. *Regul. Toxicol. Pharmacol.* 3(2, Jun): 163–171.

[66] US EPA. 1995. *Response to External Comments on New Chemical Exposure Limits in Toxic Substances Control Act '5(e) Orders*. Attachment 1. Available at: http://www.epa.gov/oppt/newchems/pubs/ncelresp.pdf (accessed June 30, 2013).

[67] US EPA. 2001. *PWB Surface Finishes - Cleaner Technologies Substitutes Assessment – Final. Appendix H – Environmental Hazard Assessment and Ecological Risk Assessment Methodology*. EPA/744-R-01-003B). Available at: http://www.epa.gov/dfe/pubs/pwb/ctsasurf/download/pdf/app-h.pdf (accessed July 14, 2013).

[68] Nabholz, J.V., Miller, P., and Zeeman, M. 1993. Environmental Risk Assessment of New Chemicals Under the Toxic Substances Control Act (TSCA) Section Five. In *Environmental Toxicology and Risk Assessment, ASTM STP 1179,* Wayne G. Landis, Jane S. Hughes, and Michael A. Lewis (Eds.). Philadelphia: American Society for Testing and Materials, pp. 40–55.

[69] Zeeman, M. 1995. EPA's Framework for Ecological Effects Assessment. In *Screening and Testing Chemicals in Commerce*. U.S. Congress, Office of Technology Assessment. OTA-BPENV-166. Washington, DC: Office of Technology Assessment. pp. 69-78. Available at: http://govinfo.library.unt.edu/ota/Ota_1/DATA/1995/9553.PDF#page=71 (accessed July 14, 2013).
[70] US EPA. 2013. *Consent Orders and Significant New Use Rules (SNURS)*. Web page last updated April 9, 2013. Available at: http://www.epa.gov/oppt/newchems/pubs/cnosnurs.htm (accessed July 19, 2013).
[71] US EPA. 2013. *New Chemicals Program - Index of Boilerplates for Orders*. Web page last updated April 3, 2013. Available at: http://www.epa.gov/oppt/newchems/pubs/boilerpl.htm (accessed July 19, 2013).
[72] US GAO. 2005. *Report to Congressional Requesters. Chemical Regulation Options Exist to Improve EPA's Ability to Assess Health Risks and Manage Its Chemical Review Program*. GAO-05-548. Available at: http://www.gao.gov/products/GAO-05-458 (accessed July 19, 2013).
[73] US EPA. 2013. *Summary of Accomplishments - New Chemicals Program Activities Through September 30, 2010*. Web page last updated April 3, 2013. Available at: http://www.epa.gov/opptintr/newchems/pubs/accomplishments.htm (accessed July 14, 2013).
[74] US EPA. 2014. *New Chemicals Program - Current Status of Section 5 Notices*. Available at: http://www.epa.gov/oppt/newchems/tools/dropstat.htm (accessed November 2, 2014).
[75] European Union. Undated. *Treaty Establishing the European Economic Community, EEC Treaty*. Available at: http://europa.eu/legislation_summaries/institutional_affairs/treaties/treaties_eec_en.htm (accessed October 11, 2013).
[76] Wilkinson, G.B. 1980. The Sixth Amendment: Toxic Substance Control in the EEC. Law Policy Int. Bus. 12 (46): 461–501.
[77] Commission of the European Communities. 1998. *Commission Working Document – Report on the Operation of Directive 67/548/EEC on the Approximation of the Laws, Regulations and Administrative Provislons Relating to the Classification, Packaging and Labelling of Dangerous Substances; Directive 88/379/EEC on the Approximation of the Laws, Regulations and Administrative Provisions Relating to the Classification, Packaging and Labelling of Dangerous Preparations; Regulation (EEC) 793/93 on the Evaluation and Control of the Risks of Existing Substances; Directive 76/769/EEC on the Approximation of the Laws, Regulations and Administrative Provisions of the Member States Relating to Restrictions on the Marketing and Use of Certain Dangerous Substances and Preparations*. Brussels,18.11.1998 - SEC(1998) 1986 final. Available at: http://aei.pitt.edu/3331/1/3331.pdf (accessed October 4, 2013).
[78] 79/831/EEC. 1979. Council Directive of 19 September 1979 amending for the sixth time Directive 67/548/EEC on the approximation of the laws, regulations and administrative provisions relating to the classification, packaging and labeling of dangerous substances. *Off. J. Euro. Commun*. L259/10-28 (Oct. 15).
[79] 81/437/EEC. 1981. *Commission Decision of 11 May 1981 laying down the criteria in accordance with which information relating to the inventory of chemical substances is supplied by the Member States to the Commission*. Available at: http://eur-lex.europa.eu/LexUriServ/LexUriServ.do?uri=CELEX:31981D0437:EN:HTML (accessed August 2, 2013).

[80] Commission of the European Communities. 2001. *WHITE PAPER: Strategy for a Future Chemicals Policy*. COM(2001) 88 final. Available at: http://eur-lex.europa.eu/LexUriServ/site/en/com/2001/com2001_0088en01.pdf (accessed August 2, 2013).

[81] Allanou, R., Hansen, B.G., and van der Bilt, Y. 1999. *Public Availability of Data on EU High Production Volume Chemicals. EUR 18996 EN*. European Commission Joint Research Centre. Available at: http://esis.jrc.ec.europa.eu/doc/datavail.pdf (accessed October 25, 2013).

[82] *Council Regulation (EEC) No 793/93 of 23 March 1993 on the Evaluation and Control of the Risks of Existing Substances.* Available at: http://eur-lex.europa.eu/smartapi/cgi/sga_doc?smartapi!celexapi!prod!CELEXnumdoc&lg=EN&numdoc=31993R0793&model=guichett (accessed October 25, 2013).

[83] European Commission Joint Research Center. Undated. *HPV-LPV Chemicals Information System.* Available at: http://esis.jrc.ec.europa.eu/index.php?PGM=hpv (accessed October 25, 2013).

[84] European Communities. 2003. *European Union Risk Assessment Report: Toluene.* Available at: http://esis.jrc.ec.europa.eu/doc/risk_assessment/REPORT/toluenereport032.pdf (accessed October 25, 2013).

[85] European Commission. 2004. *Commission Recommendation of 29 April 2004 on the Results of the Risk Evaluation and the Risk Reduction Strategies for the Substances: Acetonitrile; Acrylamide; Acrylonitrile; Acrylic acid; Butadiene; Hydrogen fluoride; Hydrogen peroxide; Methacrylic acid; Methyl methacrylate; Toluene; Trichlorobenzene.* (notified under document number C(2004) 1446) (Text with EEA relevance) (2004/394/EC). Available at: http://esis.jrc.ec.europa.eu/doc/risk_assessment/OJ_RECOMMENDATION/ojrec108883.pdf (accessed October 25, 2013).

[86] *Proposal for a Directive of the European Parliament and of the Council relating to restrictions on the marketing and use of toluene and trichlorobenzene (twenty-eighth amendment of Council Directive 76/769/EEC)/** COM/2004/0320 final - COD 2004/0111 */. Available at: http://eur-lex.europa.eu/LexUriServ/LexUriServ.do?uri=CELEX:52004PC0320:EN:HTML (accessed October 25, 2013).

[87] Fuhr, M., Merenyi, S., Bizer, K., Lewin, D., Cicborowski, G., and Kleihauer, S. 2005. *Interface Problems between EC Chemicals Law and Sector-Specific Environmental Legislation (IPPC/WFD) Environmental Research of the Federal Ministry of the Environment, Nature Conservation and Nuclear Safety.* Research Report 360 12 008. UBA-FB 000776. Available at: http://www.elni.org/fileadmin/Dokumente/Studien/2005/Interface_Problems_UBA-Texte_04-2005_05.pdf (accessed October 25, 2013).

[88] *Directive 2005/59/EC of the European Parliament and of the Council of 26 October 2005 Amending for the 28th Time Council Directive 76/769/EEC on the Approximation of the Laws, Regulations and Administrative Provisions of the Member States Relating to Restrictions on the Marketing and Use of Certain Dangerous Substances and Preparations (Toluene and Trichlorobenzene)*, Available at: http://eur-lex.europa.eu/LexUriServ/LexUriServ.do?uri=CELEX:32005L0059:en:NOT (accessed October 25, 2013).

[89] Coppock, R. 1986. *Regulating Chemical Hazards in Japan, West Germany, France, the United Kingdom, and the European Community: A Comparative Examination.* Washington, DC: National Academy Press.

[90] *Council Directive 92/32/EEC Of 30 April 1992, Amending for the Seventh Time Directive 67/548/EEC on the Approximation of the Laws, Regulations and Administrative Provisions Relating to the Classification, Packaging and Labeling of Dangerous Substances*. Available at: http://eur-lex.europa.eu/LexUriServ/LexUriServ.do?uri=OJ:L:1992:154:0001:0029:EN:PDF (accessed September 2013).

[91] European Commission. 2009. *ELINCS*. Available at: http://esis.jrc.ec.europa.eu/index.php?PGM=eli (accessed October 25, 2013).

[92] European Commission. 2007. *REACH in Brief*. Available at: http://ec.europa.eu/environment/chemicals/reach/pdf/publications/2007_02_reach_in_brief.pdf (accessed November 1, 2013) and also Commission of the European Communities, 2001. *WHITE PAPER: Strategy for a future Chemicals Policy*. COM(2001) 88 final. Available at: http://eur-lex.europa.eu/LexUriServ/site/en/com/2001/com2001_0088en01.pdf (accessed August 2, 2013).

[93] Anonymous. 2011. *Regulatory Framework for the Management of Chemicals (REACH), European Chemicals Agency*. Web page last updated February 23, 2011. Available at: http://europa.eu/legislation_summaries/internal_market/single_market_for_goods/chemical_products/l21282_en.htm (accessed November 11, 2013).

[94] *Regulation (EC) No 1907/2006 of the European Parliament and of the Council of 18 December 2006 concerning the Registration, Evaluation, Authorisation and Restriction of Chemicals (REACH), Establishing a European Chemicals Agency, Amending Directive 1999/45/EC and Repealing Council Regulation (EEC) No 793/93 and Commission Regulation (EC) No 1488/94 as well as Council Directive 76/769/EEC and Commission Directives 91/155/EEC, 93/67/EEC, 93/105/EC and 2000/21/EC*. Available at: http://eur-lex.europa.eu/LexUriServ/LexUriServ.do?uri=CELEX:32006R1907:EN:NOT (accessed November 11, 2013).

[95] ECHA. Undated. *Guidance on Information Requirements and Chemical Safety Assessment*. Available at: http://echa.europa.eu/web/guest/guidance-documents/guidance-on-information-requirements-and-chemical-safety-assessment (accessed November 29, 2013).]

[96] ECHA. Undated. *Animal testing under REACH*. Available at: http://echa.europa.eu/chemicals-in-our-life/animal-testing-under-reach (accessed November 1, 2013).

[97] ECHA. 2011. *Guidance on Information Requirements and Chemical Safety Assessment. Part B: Hazard Assessment*. ECHA-11-G-16-EN. Available at: http://echa.europa.eu/documents/10162/13643/information_requirements_part_b_en.pdf (accessed November 29, 2013).

[98] European Chemicals Agency. 2009. *Guidance in a Nutshell - Chemical Safety Assessment*. Available at: http://echa.europa.eu/documents/10162/13632/nutshell_guidance_csa_en.pdf (accessed November 29, 2013).

[99] ECHA. 2012. *Guidance on Information Requirements and Chemical Safety Assessment. Part D: Exposure Scenario Building*. Version 1.2. Available at: http://echa.europa.eu/documents/10162/13632/information_requirements_part_d_en.pdf (accessed November 29, 2013).

[100] ECHA. 2012. *Guidance on Information Requirements and Chemical Safety Assessment. Chapter R.14: Occupational Exposure Estimation*. ECHA-2010-G-09-EN. Available at: http://echa.europa.eu/documents/10162/13632/information_requirements_r14_en.pdf (accessed November 29, 2013).

[101] ECHA. 2012. *Guidance on Information Requirements and Chemical Safety Assessment. Chapter R.15: Consumer Exposure Estimation.* ECHA-10-G-03-EN. Available at: http://echa.europa.eu/documents/10162/13632/information_requirements_r15_en.pdf (accessed November 29, 2013).

[102] Delmaar, J.E., Park, M.V.D.Z., and van Engelen, J.G.M. 2005. *ConsExpo 4.0 Consumer Exposure and Uptake Models Program Manual.* RIVM report 320104004/2005. Available at: http://www.rivm.nl/dsresource?objectid=rivmp:13097&type=org&disposition=inline&ns_nc=1 (accessed November 29, 2013).

[103] Oppl, R., Kalberlah, F., Evans, P.G., and Van Hemmen, J.J. 2003. A Toolkit for Dermal Risk Assessment and Management: An Overview. *Ann. Occup. Hyg.* 47(8): 629–640.

[104] Fransmen, W. et al. 2013. *Development of a Mechanistic Model for the Advanced REACH Tool (ART) - Version 1.5.* Available at: https://www.advancedreach-tool.com/assets-1.5.12110.3/doc/ART%20Mechanistic%20model%20report_v1_5_20130118.pdf (accessed November 29, 2013).

[105] European Commission Joint Research Center. Undated. *GExFRAME v1.1 - Introduction.* Available at: http://gexframe.jrc.ec.europa.eu/GExFRAME/Intro.aspx (accessed November 29, 2013).

[106] ECHA. 2012. *Guidance on Information Requirements and Chemical Safety Assessment. Chapter R.16: Environmental Exposure Estimation.* ECHA-10-G-06-ENN. Available at: http://echa.europa.eu/documents/10162/13632/information_requirements_r16_en.pdf (accessed November 29, 2013).

[107] CEM – Netherlands Center for Environmental Modeling. Undated. *EU TGD Spreadsheet.* Available at: http://cem-nl.eu/eutgd.html (accessed November 29, 2013).

[108] ECHA. 2011. *Background Document to the Decision of the Executive Director of ECHA, ED/32/2011 Selection Criteria to Prioritise Substances for Substance Evaluation (2011 CoRAP Selection Criteria).* Available at: http://echa.europa.eu/documents/10162/13628/background_doc_criteria_ed_32_2011_en.pdf (accessed November 10, 2013).

[109] ECHA. 2012. *Fact Sheet: Substance Evaluation. ECHA-11-FS-03-EN.* Available at: http://echa.europa.eu/documents/10162/13628/fs_substance_evaluation_en.pdf (accessed November 10, 2013.)

[110] ECHA. Undated. *The Candidate List.* Available at: http://echa.europa.eu/regulations/reach/authorisation/the-candidate-list (accessed November 10, 2013).

[111] ECHA. Undated. *Summary of Obligations Resulting from Inclusion in the Candidate List of Substances of Very High Concern for Authorisation.* Available at: http://echa.europa.eu/candidate-list-obligations (accessed November 10, 2013).

[112] ECHA. 2011. *Guidance on the preparation of an application for authorization. Version 1. January.* Available at: http://echa.europa.eu/documents/10162/13643/authorisation_application_en.pdf (accessed November 10, 2013).

[113] ECHA. 2013. *Applications for Authorisation.* Available at: http://echa.europa.eu/web/guest/addressing-chemicals-of-concern/authorisation/applications-for-authorisation (accessed November 13, 2013).

[114] ECHA. 2010. *Procedure on Restrictions.* Available at: http://echa.europa.eu/documents/10162/13607/procedure_restrictions-v01_en.pdf (accessed December 6, 2013).

[115] ECHA. Undated. *Restriction Procedure*. Available at: http://echa.europa.eu/web/guest/regulations/reach/restrictions/restriction-procedure (accessed December 6, 2013.

[116] European Commission. 2004. *European Union Risk Assessment Report - 1,4-Dichlorobenzene. CAS No: 106-46-7. EINECS No: 203-400-5. Risk Assessment.* Available at: http://echa.europa.eu/documents/10162/fb7bf6b4-7831-4c3b-87b3-5acd493ce597 (accessed December 8, 2013).

[117] European Commission. 2008. *Commission Communication on the Results of the Risk Evaluation and Risk Reduction Strategies for the Substances: Piperazine; Cyclohexane; Methylenediphenyl diisocyanate; But-2yne-1,4-diol; Methyloxirane; Aniline; 2-Ethylhexylacrylate; 1,4-Dichlorobenzene; 3,5-dinitro-2,6-dimethyl-4-tert-butylacetophenone; Di-(2ethylhexyl)phthalate; Phenol; 5-tert-butyl-2,4,6-trinitro-m-xylene.* (2008/C 34/01) Available at: http://esis.jrc.ec.europa.eu/doc/risk_assessment/OJ_RECOMMENDATION/ojrec106467.pdf (accessed December 8, 2013.

[118] ECHA. 2013. *Committee for Risk Assessment (RAC)/Committee for Socio-Economic Analysis (SEAC) - Background Document to the Opinion on the Annex XV Dossier Proposing Restrictions on 1,4-Dichlorobenzene.* ECHA/RAC/RES-O-0000003486-69-01/F. ECHA/SEAC/RES-O-0000003486-69-02/F. Available at: http://echa.europa.eu/documents/10162/7b884aff-2852-40c3-8e7a-642ca92ca425 (accessed December 8, 2013)]

[119] European Commission. 2013. *Notification to World Trade Organisation Committee on Technical Barriers to Trade.* G/TBT/N/EU/145. 12 September. Available at: http://www.wtocenter.org.tw/SmartKMS/fileviewer?id=135789 (accessed December 19, 2013).

[120] Council of the European Union. 2014. *Commission Regulation (EU) No …/.. of XXX amending Annex XVII to Regulation (EC) No 1907/2006 of the European Parliament and of the Council on the Registration, Evaluation, Authorisation and Restriction of Chemicals ("REACH") as Regards 1,4-Dichlorobenzene.* 6 January. Available at: http://register.consilium.europa.eu/doc/srv?l=EN&t=PDF&f=ST+5208+2014+INIT (accessed March 8, 2014).

[121] Song, Y. 2002. The Chinese Environmental Lawmaking Framework. *Chinese J. Int. Law.* 1 (1): 225–233.

[122] Beyer, S. 2006. Environmental Law and Policy in the People's Republic of China. *Chinese J. Int. Law.* 5(1): 185–211.

[123] *Law of the People's Republic of China on Promotion of Cleaner Production (Order of the President No.72).* (Adopted at the 28th Meeting of the Standing Committee of the Ninth National People's Congress on June 29, 2002). Available at: http://www.gov.cn/english/laws/2005-10/08/content_75059.htm (accessed July 26, 2013).

[124] Chemical Registration Center of State Environmental Protection Administration. 2004. *Technical Document on Reporting for the Inventory of Existing Chemical Substances in China.* Available at: http://www.crc-mep.org.cn/newchem/ebcsl.htm (accessed March 15, 2014).

[125] Chemical Inspection and Regulation Service. 2014. *Toxic Chemicals Restricted To Be Imported/Exported.* Updated January. Available at: http://www.cirs-reach.com/China_Chemical_Regulation/Registration_of_import_export_of_toxic_chemicals_in_China.html (accessed March 13, 2014).

[126] *Detailed Rules on Implementing the Registration of the Environmental Management on the First Import of Chemicals and the Import and Export of Toxic Chemicals.* Available at: http://www.crc-mep.org.cn/English/e022.htm (accessed May 2013).

[127] MEP. 2012. *List of Toxic Chemicals Severely Restricted to Be Imported into or Exported from China*. Available at: http://www.mep.gov.cn/gkml/hbb/bgg/201112/W020111231403819465014.pdf (accessed July 27, 2013).
[128] Chemical Inspection and Regulation Service. 2014. *China Revises the List of Toxic Chemicals Restricted to Be Imported into or Exported from China*. January 3. Available at: http://www.cirs-reach.com/news/China_Revises_the_List_of_Toxic_Chemicals_Restricted_to_Be_Imported_into_or_Exported_from_China.html (accessed March 13, 2014).
[129] CRC-MEP. 2006. *Guide for Registration of Import/Export of Toxic Chemicals*. Available at: http://www.crc-mep.org.cn/en/A107/A107_S1.aspx (accessed March 22, 2014).
[130] Chemical Inspection and Regulation Service. 2013. *Regulations on Safe Management of Hazardous Chemicals in China - Decree 591*. Available at: http://www.cirs-reach.com/China_Chemical_Regulation/Regulations_on_Safe_Management_of_Hazardous_Chemicals_in_China.html (accessed March 22, 2014).
[131] Burgess, A. and Sullivan, D.E. 2011. Decree 591: A new era of chemical control legislation in China. Presented at *ChemCON Asia 2011*. June 28.
[132] Chemical Watch. 2012. *China Issues Measures for Environmentally Hazardous Chemicals*. October 18. Available at: http://chemicalwatch.com/12658/china-issues-measures-for-environmentally-hazardous-chemicals (accessed March 22, 2014).
[133] Chemical Inspection and Regulation Service. 2013. *China SAWS Publishes the Revised Catalogue of Hazardous Chemicals for Consultations*. Available at: http://www.cirs-reach.com/news/China_SAWS_Publishes_the_Revised_Catalogue%20of%20Hazardous_Chemicals_for_Consultations.html (accessed March 22, 2014).
[134] Chynoweth, E. 2014. China's HazChem Catalogue expected shortly; Final batch of substances being discussed next week. *Chemical Watch*. October 30. Available at: http://chemicalwatch.com/21777/chinas-hazchem-catalogue-expected-shortly (accessed November 2, 2014).
[135] Ning, D., Yan-ying, B., Xiu-ling, Y., Jie, Y., and Dan-na, S. 2009. Analysis on cleaner production policy and its results in China. In *IWACP: Second International Workshop – Advances in Cleaner Production*. São Paulo, Brazil, May 20–22, 2009. Available at: http://www.advancesincleanerproduction.net/second/files/sessoes/4a/1/D.%20Ning%20-%20Relatorio%20-%204A-1.pdf (accessed March 15, 2014).
[136] Dan-na, S., Ning, D., Yan-ying, B, and Jie, Y. 2011. Exploration and practice on the compulsory cleaner production auditing of key enterprise in China. In *IWACP: Third International Workshop – Advances in Cleaner Production*. São Paulo, Brazil, May 18–20, 2011. Available at: http://www.advancesincleanerproduction.net/third/files/sessoes/4B/5/Song_Danna%20-%20Paper%20-%204B5.pdf (accessed March 15, 2014).
[137] Ministry of Environmental Protection. 2013. *Prevention and Control of Environmental Risks of Chemicals "Twelfth Five Year Plan" [translation]*. Available at: http://www.mep.gov.cn/gkml/hbb/bwj/201302/W020130220539067366659.pdf (accessed March 22, 2014).
[138] China Council for International Cooperation on Environment and Development. 2013. *MEP releases the "12th Five-Year Plan for Prevention and Control of Environmental Risk of Chemicals"*. Available at: http://www.cciced.net/encciced/newscenter/latestnews/201302/t20130225_248445.html (accessed March 22, 2014).

[139] *Provisions on the Environmental Administration of New Chemical Substances (SEPA Order number 17)* (available at: http://www.crc-mep.org.cn/newchem/enewchem.htm (accessed July 26, 2013).

[140] Chemical Registration Center of SEPA. 2004. *Guidance for New Chemical Substance Notification (First version)*. October. Available at: http://english.mep.gov.cn/special_reports/chemical_regulation/newchemicals/200712/P020071217518150802043.pdf (accessed July 27, 2013).

[141] Yingxin, G. 2007. *Environmental Management on Chemicals in China. Presentation March 30.* Chemical Registration Center of SEPA. Available at: https://www.env.go.jp/chemi/reach/second/gao.pdf (accessed July 27, 2013).

[142] Chemical Registration Center of the Ministry of Environmental Protection. 2010. *Inventory of Existing Chemical Substances in China, 2010.* Available at: http://www.crc-mep.org.cn/en/M006/M006_C1.aspx (accessed March 15, 2014).

[143] Jing, Y. 2011. China – Significant changes in chemical control legislation – past, present, and future. Chemical Registration Center of MEP. Presented at *ChemConAsia* 2011. June 28, 2011. Available at: http://www.lawbc.com/share/chemcon2011/00079429.pdf (accessed July 27, 2013).

[144] *Environmental Administration of New Chemical Substances.* Available at: http://www.gov.cn/flfg/2010-02/04/content_1528001.htm (accessed March 22, 2014).

[145] China Ministry of Environmental Protection. 2010. *Notice No. 2010-124 – Notice of Guidance for New Chemical Substance Notification and Registration.* Issued November 15. (unofficial translation).

[146] Chemical Inspection and Regulation Service. Undated. *Guidance Documents for New Chemical Substance Notification in China (China REACH) – Non Official.* Available at: http://www.cirs-reach.com/China_Chemical_Regulation/Guidance_New_Chemical_Registration_China.pdf (accessed March 13, 2014).

[147] Ma, G. 2014. *China: New Substance Notification. Chemical Watch Global Business Briefing,* March 2014. Available at: http://chemicalwatch.com/18661/china-new-substance-notification. (accessed March 10, 2014).

[148] Chemistry Industry Association of Canada. Undated. *A Brief History of Responsible Care.* Available at: http://www.canadianchemistry.ca/responsible_care/index.php/en/responsible-care-history (accessed December 26, 2013).

[149] International Council of Chemical Associations. 2012. *ICCA RESPONSIBLE CARE® PROGRESS REPORT - Growing Our Future.* Available at: http://www.icca-chem.org/ICCADocs/RC%20annual%20report.pdf (accessed December 26, 2013).

[150] Chemical Manufacturers' Association. 1991. Responsible Care: A Public Commitment. As cited in: King, A. A., and Lenox, M.J. 2000. Industry self-regulation without sanctions: The chemical industry's responsible care program. *Acad. Manage. J.* 43(4): 698–716.

[151] ACC. 2013. *Responsible Care® Guiding Principles.* Available at: http://responsiblecare.americanchemistry.com/Responsible-Care-Program-Elements/Guiding-Principles/PDF-Responsible-Care-Guiding-Principles.pdf (accessed December 26, 2013).

[152] CMA. 1992. *Product Stewardship Code of Management Practices.*

[153] ACC. 2013. *Responsible Care®: Milestones and Achievements.* Available at: http://responsiblecare.americanchemistry.com/Home-Page-Content/Responsible-Care-Timeline.pdf (accessed December 26, 2013).

[154] ACC. 2013. *Responsible Care® Product Safety Code of Management Practices.* Available at: http://responsiblecare.americanchemistry.com/Responsible-Care-Program-Elements/Product-Safety-Code/Responsible-Care-Product-Safety-Code-PDF.pdf (accessed December 26, 2013).
[155] Mazurek, J. 1998. *The Use of Unilateral Agreements in the United States: The Responsible Care Initiative.* OECD Working Party on Economic and Environmental Policy Integration. ENV/EPOC/GEEI(98)25/FINAL. Available at: http://search.oecd.org/officialdocuments/displaydocumentpdf/?doclanguage=en&cote=ENV/EPOC/GEEI(98)25/FINAL (accessed December 27, 2013).
[156] ACC. 2013. *Responsible Care® Performance Results.* Available at: http://responsiblecare.americanchemistry.com/Performance-Results (accessed December 26, 2013).
[157] International Council of Chemical Associations. 2012. *ICCA Responsible Care® Progress Report - Third Session of the International Conference on Chemicals Management September 17–21, 2012.* Available at: http://www.icca-chem.org/ICCADocs/ICCA%20Progress%20Report.pdf (accessed December 26, 2013).
[158] ACC. 2013. *Responsible Care Performance Results – Safety.* Available at: http://responsiblecare.americanchemistry.com/Performance-Results/Safety (accessed December 28, 2013).
[159] ACC. 2013. *Responsible Care Performance Results – Environment.* Available at: http://responsiblecare.americanchemistry.com/Performance-Results/Environment (accessed December 27, 2013).
[160] King, A.A. and Lenox, M.J. 2000. Industry Self-Regulation without Sanctions: The Chemical Industry's Responsible Care Program. *Acad. Manage. J.* 43(4): 698–716.
[161] Gamper-Rabindran, S. and Finger, S.R. 2013. Does industry self-regulation reduce pollution? Responsible Care in the chemical industry. *J. Regul. Econ.* 43;1–30.
[162] Cefic. 2013. *Responsible Care Performance Reporting.* Available at: http://www.cefic.org/Responsible-Care/Performance (accessed December 28, 2013).
[163] Cefic. 2013. *Responsible Care – Environment/Air.* Available at: http://www.cefic.org/Responsible-Care/Performance/Environment/(accessed December 28, 2013).
[164] CPCIA. 2008. *China Petroleum and Chemical Industry Association (CPCIA) Responsible Care Annual Report 2008.* Available at: http://www.cpcia.org.cn/Picture/edit/2010/2010092810113264.pdf (accessed December 26, 2013).
[165] CPCIF. 2011. *Proposal on Promotion of "Responsible Care" in Petrochemical Industry.* Available at: http://www.cpcia.org.cn/Picture/edit/2011/2011051314592183.pdf (accessed December 26, 2013).
[166] CPCIF. 2011. *CPCIF Annual Report 2010.* Available at: http://www.cpcia.org.cn/Picture/edit/2011/2011040115220533.pdf (accessed December 28, 2013).
[167] Roberts, G. 2013. *Chemical Watch Briefing: Promoting Safety in China.* May 24. Available at: http://chemicalwatch.com/14973/promoting-safety-in-china?q=%22responsible%20care%22 (accessed December 26, 2013).
[168] ICCA. 2013. *Responsible Care Members - Hong Kong SAR, China.* Available at: http://www.icca-chem.org/en/Home/Responsible-care/Responsible-Care-Members/?country=bj (accessed December 28, 2013).
[169] Association of International Chemical Manufacturers. 2009. *Responsible Care Committee.* Available at: http://www.aicm.cn/wyh1.aspx?id=3 (accessed December 26, 2013).

[170] International Council of Chemical Associations. 2012. ICCA Responsible Care® Progress Report. In *Third Session of the International Conference on Chemicals Management*, September 17–21, 2012. Available at: http://www.icca-chem.org/ICCADocs/ICCA%20Progress%20Report.pdf (accessed December 26, 2013).

[171] Environmental Defense and DuPont. 2007. *Nano Risk Framework*. 21 June. www.NanoRiskFramework.com (accessed June 27, 2007).

[172] Environmental Defense and DuPont. 2007. *Nanomaterial Risk Assessment Worksheet: DuPont™ Light Stabilizer for Use as a Polymer Additive.* http://www.environmentaldefense.org/documents/6552_TiO2_Summary.pdf (accessed June 27, 2007).

[173] Warheit, D.B. 2013. How to measure hazards/risks following exposures to nanoscale or pigment-grade titanium dioxide particles. *Toxicol. Lett.* 220(2): 193–204.

[174] SC Johnson. 2013. *Our Greenlist™ Process*. Available at: http://www.scjohnson.com/en/commitment/focus-on/greener-products/greenlist.aspx (accessed October 6, 2013).

[175] Layton, L. 2011. Wal-Mart bypasses federal regulators to ban controversial flame retardant. *The Washington Post*. Available at: http://www.washingtonpost.com/wp-dyn/content/article/2011/02/25/AR2011022502977_pf.html (accessed October 6, 2013).

[176] US EPA. 1983. *Regulation of Genetically Engineered Substances under TSCA*. Draft.

[177] US EPA. 1997. *Points to Consider in the Preparation of TSCA Biotechnology Submissions for Microorganisms*. June. Available at: http://www.epa.gov/oppt/biotech/pubs/pdf/ptcbio.pdf (accessed April 18, 2014).

[178] *Bill Summary & Status - 113th Congress (2013–2014) S.1009*. Available at: http://thomas.loc.gov/cgi-bin/bdquery/z?d113:s.1009: (accessed November 2, 2014).

[179] Bergeson and Campbell (law firm). 2014. *Regulatory Developments Summary of Changes in the Chemical Safety Improvement Act(CSIA, S. 1009) and the Boxer TSCA Version Released September 18, 2014 (September 25, 2014)*. Available at: http://www.lawbc.com/regulatory-developments/entry/summary-of-changes-in-the-chemical-safety-improvement-actcsia-s.-1009-and-t (accessed November 2, 2014).

[180] Kumar, D. 2014. Chance of 2014 TSCA reform in US Congress dwindling - Industry sanguine about 2015 passage. *Chemical Watch*. August 26. Available at: http://chemicalwatch.com/20964 (accessed November 2, 2014).

4

Case Studies

The preceding chapters used brief case studies to illustrate key points. This chapter examines the life cycle of four substances in greater detail. Three of these substances, orthonitrochlorobenzene, 1,4-dichlorobenzene, and hexachlorobenzene, share a basic chlorinated benzene structure. The degree of chlorination and the presence of other functional groups affect their properties, usage, fate and transport in the environment, and (eco)toxicity. These substances also differ in their uses, which influences the potential for exposure. The remaining case study examines microbeads, a product whose size determines in part its life cycle.

4.1 Orthonitrochlorobenzene

This story begins from an environmental perspective in the late 1950s. It is worth a close look because the story illustrates the efforts of initial environmental regulations to catch up to the consequences of the Industrial Revolution and, more broadly, the evanescence of some environmental issues as the marketplace evolves.

4.1.1 Introduction

Few environmental laws or regulations existed in the United States in the 1950s. One of those laws, the Federal Water Pollution Control Act (FWPCA) of 1948, largely referred to planning and coordination activities intended to limit pollution in interstate waterways. Not until Congress amended the FWPCA in 1956 did the federal government actually have the authority to regulate the quality of water that flowed across state lines. Just as important, the 1956 amendments provided for federal grants to construct municipal wastewater treatment facilities.

In order to establish trends in contamination levels in major river systems—and ideally to document progress in cleaning up streams as a result of wastewater treatment plants funded though the FWPCA—the US Public Health Service established a Water Quality Network. The initial network provided water quality data at 50 surface water monitoring stations around the country. It's no exaggeration to say that the establishment of this network catalyzed the development of environmental analytical chemistry and toxicology in the United States.

The scientists working on this project lacked most of the tools we now take for granted. Collecting and analyzing samples required extraordinary effort. The US Public Health Service devised means to collect samples at each station on the order of 3,000 to 4,000 gallons of water each and then passed each of those samples through a carbon filter to isolate organic contaminants. Analytical chemists separated the organic residues into fractions based on the solubility of chemical families in various solvents, then used infrared spectroscopy and the melting points of solids to identify individual compounds [1,2].

This painstaking work produced startling results. Researchers found orthonitrochlorobenzene (ONCB) in water samples collected over 900 miles of the Mississippi River in 1958. They traced the compound to a single apparent source near the city of St. Louis, Missouri, a discharge from a factory that manufactured paranitrochlorobenzene and produced ONCB as an "unusable byproduct" [3]. (This source was apparently a facility located in Sauget, Illinois [4], across the river from St. Louis.)

Samples from the monitoring station approximately 100 miles downstream of the apparent source, at Cape Girardeau, Missouri, typically contained the highest concentrations of ONCB detected in the US Public Health Service study. All of the samples collected each month from this station during 1958 contained ONCB, at an average concentration of 21 parts per billion (ppb). Researchers from the US Public Health Service conservatively estimated that 22,600 pounds (approximately 10,300 kg) of ONCB flowed past this monitoring point every 24 hours [3].

The data from the monitoring station near Cape Girardeau pointed to another problem. The city of Cape Girardeau obtained its drinking water from the Mississippi River and the raw water intake contained ONCB. When the US Public Health Service analyzed samples of treated water, the data showed that drinking water treatment did not reduce the concentration of ONCB [3]. As a result the residents of Cape Girardeau ingested ONCB with their drinking water; a sample of treated water collected during one week in 1958 contained 30 ppb ONCB. These data were among the first to quantify the exposure to an anthropogenic chemical from an environmental source in the United States.

The scientists from the US Public Health Service also lacked the risk-assessment tools we rely on now to understand the implications of exposure to chemical substances. Based on the toxicological data available at the time, they estimated that the probable lethal dose of ONCB to a human was between 5 and 50 milligrams per kilogram (mg/kg) of body weight, or "between 7 drops and 1 teaspoonful for 150 lb.-man [*sic*]. A clinically significant illness would be expected to result from 1/10 the probable lethal dose" [3, Appendix, p. 2]. They further calculated that the estimated daily dose would allow for a 467-fold factor of safety but concluded that, given the uncertainties, "it does not seem reasonable to have such an additive in drinking water in any concentration when it is possible to eliminate it at the source" (p. 7). An addendum to the US Public Health Service report on the study of Cape Girardeau drinking water written in 1959 notes that [3, p. 7]: "Once this chemical was identified, the industry that was most likely to be discharging it was contacted on an informal basis ... the waste stream was immediately taken out of the river and the chemical has not reappeared." A decade later, this description of ONCB in the Mississippi River helped to catalyze the development of new regulations. In its 1971 report *Toxic Substances*, the President's Council on Environmental Quality cited the data from this study to illustrate the need for TSCA [5]. With that introduction, let's look at the life cycle of this chemical, which is variously referred to as ONCB, CAS No. 88-73-3, and 1-chloro-2-nitrobenzene.

4.1.2 Physicochemical Characteristics

Table 4.1 summarizes fundamental information about critical physicochemical properties of ONCB and the other chlorinated benzenes discussed in this chapter.

4.1.3 Manufacture and Use

Manufacturers commonly produce nitrochlorobenzene compounds by nitrating chlorobenzene under acidic conditions. The reaction produces a mixture of isomers comprising by weight 34% ortho, 65% para, and 1% meta nitrochlorobenzenes. The isomers are separated by crystallization and distillation [25].

The early manufacturing process was quite simple and the consequences of worker exposure notable [26, p. 19]:

> In 1923, an English chemical firm began manufacturing 2-chloronitrobenzene and 4-chloronitrobenzene. The nitration of chlorobenzene was carried out in a large tank, and the resulting hot oil was cooled in trays. 4-Chloronitrobenzene was crystallized, and the supernatant, 2-chloronitrobenzene, was run off into drums. 4-Chloronitrobenzene was shoveled into a centrifuge for drying and then placed in casks. Workers generally required hospitalization after about 3 days of exposure. Symptoms included a slate gray appearance, headache, and dyspnea on exertion. Blood serum was often a port wine color, and erythrocytes were large and occasionally deformed. Workers received treatments of atropine, ether, pituitary extract, and a coffee enema. Despite this, all recovered over a period of several weeks.

By 1976, production utilized more sophisticated batch processes [4]. Modern production processes operate in a closed continuous flow system [27].

Production data are scarce, in part due to the limited number of manufacturers and the resulting business need to keep specific production information

TABLE 4.1

Properties of Chlorinated Benzenes

Compound (CAS no.)	Solubility (mg/L)	Henry's Law Constant (atm-cu m/mole)	Half Life	Log K_{aw} [b]	Log K_{ow}	Log K_{oc}	BCF, Fish
1-chloro, 2-nitrobenzene (88-73-3)	441 mg/L at 25 deg C [6]	9.3×10^{-6} [7]	Water: Not readily biodegradable [8] Air: 94 days (calc.) [9]	–3.4	2.52 [10]	2.50 [9]	Carp: 7 – 22.3 [11]
1,4- dichorobenzene (106-46-7)	80.0 mg/L [12]	2.41×10^{-3} at 25C [13]	Water: 900 hours (calc.) Air: 802 hours (calc.) [14]	–1.0	3.44 [15]	2.60 [16]	Rainbow trout: 370 – 720 [17]
Hexachlorobenzene (118-74-1)	0.006 mg/L at 20 C [18]	1.7×10^{-3} at 25C [19][a]	Surface water: 2.7–5.7 Groundwater: 5.3–11.4 Air: 0.63–6.28, depending on temperature [20, 21]	–1.2	5.73 [22]	2.51 [23]	Fish: Up to 21,900 [24]

[a] The values reported in the literature for the Henry's law constant of HCB vary over four orders of magnitude. The experimentally derived value presented here is consistent with the reference value used in EpiSuite 4.11.

[b] Calculated from the values of the Henry's law constant shown in this table.

confidential. The United States International Trade Commission, which published data on the manufacture and import of synthetic organic chemicals through 1994, limited its reporting on ONCB in later years to avoid disclosure of individual company operations [28]. Similarly, quantitative information is not publicly available from the 2012 Chemical Data Reporting (CDR) on the production and use of chemicals manufactured or imported into the United States, as only a single US company manufactured or imported ONCB in an amount greater than 25,000 pounds per year [29]. Despite such strictures, some information on production rates in the United States, European Union, and China are available as shown in Table 4.2.

TABLE 4.2

Global Production of ONCB

	Approximate Annual Production (Metric Tonnes)				
Year	Global	United States	European Union	China	Reference
1960		11,131			[4]
1961		7,791			[4]
1963		8,563			[4]
1964		9,112			[4]
1965		12,832			[4]
1966		14,617			[4]
1967		15,543			[4]
1968		>6,633			[4]
1969		>9,249			[4]
1977		>4,990			[9]
1981		<16,300			[9]
1985		13,600	>20,400[a]		[30]
1986		4,536–22,680			[9]
1990		22,680–45,359		15,800	[9]
1993		19,099			[25]
1994		22,680–45,359			[9]
1995	111,800	17,000	25,000[b]	36,960	[27, 31]
1996				38,050	[31]
1997				40,400	[31]
1998		22,680–45,359		50,200	[9, 31]
2002		4,536–22,680			[9]
2006	16,200			10,500	[31]
2012		0? [c]			[29]
2014				>95,200?[d]	[32]

[a] Production reported for West Germany only.
[b] Western Europe only.
[c] CDR data indicate import only.
[d] Total production 280,000 metric tonnes per year para and orthonitrochlorobenzene by one manufacturer alone, Anhui Bayi Chemical Industry Co., Ltd.; estimated production of ONCB as 34% of total.

Given the gaps in knowledge and the vagueness of some of the values, it is difficult to interpret these production data with respect to trends in the manufacture and usage of ONCB. Reports by various authorities suggest a gradual decrease in production since 2000 and a general shift in production from the United States and European Union to China; the reported production by one manufacturer in China circa 2014, which is indicated in Table 4.2, may suggest a dramatic increase in demand or may be an anomaly.

Although the Public Health Service characterized ONCB as an "unusable byproduct" of manufacturing other substances in 1958, ONCB has found many uses as a chemical intermediate. As early as the 1960s, it was used to produce 2-nitroaniline, 2-chloroaniline, 2-anisidine, 2-phenetidine, and 2-aminophenol [4]. By 1995 the primary worldwide uses of ONCB were as follows [27]:

- Production of 2-nitroaniline (31%), an intermediate in pesticides production
- Synthesis of dichlorobenzidine (26%), 2-nitroanisole (23%), and 2-chloroaniline (8%), used in dyes and pigments
- Other intermediates (12%) in the production of dyes and pigments, pharmaceuticals, and pesticides

Circa 2004, a US-based manufacturer noted that ONCB was then used to manufacture dyes and pigments, polymer additives, veterinary pharmaceuticals, and water-treatment chemicals [33].

4.1.4 Fate and Transport in the Environment

The data in Table 4.1 indicate that ONCB has limited water solubility and volatility. ONCB does not appear to biodegrade readily. The Level III fugacity model within EpiSuite 4.11 predicts the environmental distribution and half-lives in various media shown in Figures 4.1 and 4.2. (As discussed in more detail in Chapter 2, these model results do not serve as an absolute predictor of the fate and transport of a substance, but rather indicate in general terms the tendency of a substance to partition between compartments and biodegrade.)

As described in Chapter 2, the Stockholm Convention on Persistent Organic Pollutants defines a chemical as persistent if its half-life in water is greater than two months (1,440 hours), or the half-life in soil or sediment is greater than six months (4,230 hours); additionally, a chemical is considered to have the potential for long-range transport in air if its half-life (in air) is greater than two days. A high potential for bioaccumulation is indicated by a bioconcentration factor (BCF) or bioaccumulation factor (BAF) in aquatic species greater than 5,000 or alternatively by a log K_{ow} value above 5 [34]. A comparison of these criteria to the data in Table 4.1 and the estimates illustrated in Figure 4.1 suggest that ONCB would not persist in water or soil but

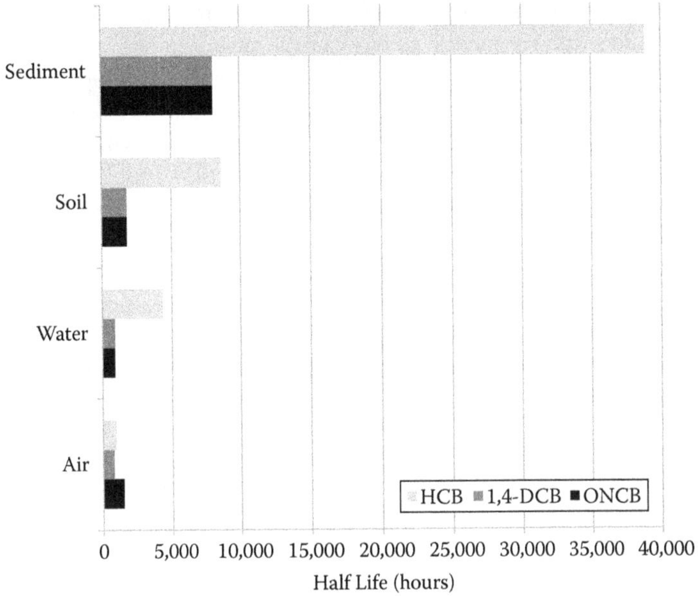

FIGURE 4.1
Half-lives (as predicted by EpiSuite 4.11).

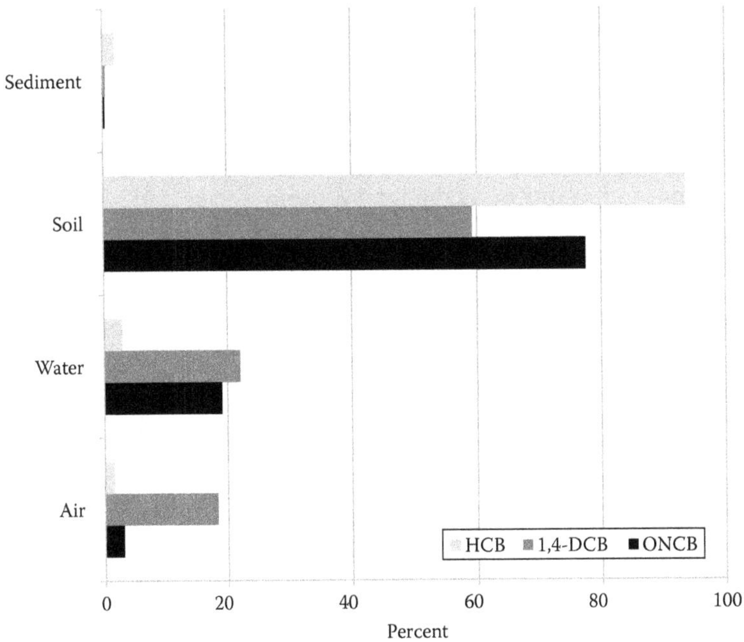

FIGURE 4.2
Distribution between environmental compartments (as predicted by EpiSuite™ 4.11).

has the potential to persist in sediment and to be transported long distances by air. It would not be expected to bioaccumulate.

The Organization for Economic Cooperation and Development (OECD) has developed another way to assess the potential for persistence and long-range transport. The *OECD Pov and LRTP Screening Tool* utilizes the Henry's law constant (also referred to as the air–water partition coefficient, K_{aw}), K_{oc}, and half-lives in environmental media to calculate three factors [35,36]:

- *Overall persistence (Pov):* Characteristic time (days) for disappearance of a chemical, based on the modeled partitioning between media (using a Level III fugacity model) and the degradation rate in each medium. The benchmark value, 195 days, is based on the persistence of α-HCH.
- *Characteristic travel distance (CTD):* The distance (km) at which the concentration of a substance has decreased to 37% of the original concentration, assuming that the substance is transported by a constant flow of water (ocean speed 0.02 m/s) or air (4 m/s). The benchmark value, 5,097 km, is based on the model results for the reference chemical PCB-28.
- *Transport efficiency (TE):* Estimated percentage of a substance that is deposited to surface media after transport away from the point of release; this metric describes the potential for atmospheric transport and deposition of a compound in a remote region. The benchmark value for TE is 2.248% based on the model results for the reference chemical PCB-28.

Modeled results above the benchmark values suggest—subject to the limitations of the model—the potential for a substance to be persistent and travel long distances.

Table 4.3 indicates the values derived from the *OECD Pov and LRTP Screening Tool.* The results suggest that ONCB is, relatively speaking, not very persistent but has some potential to be transported long distances, particularly when emitted to air.

As to the consequences of exposure to this substance in the environment, let's briefly review the toxicological and (eco)toxicological data.

4.1.5 Overview of (Eco)Toxicological Concerns

ONCB is considered to be moderately toxic to humans upon acute exposure. Regardless of the route of exposure, cyanotic appearance is the predominant symptom observed in toxicity tests. Repeated exposure to sufficient concentrations of ONCB may cause methemoglobinemia [27]. This blood disorder, in which the body produces an abnormal level of methemoglobin and cannot effectively transport oxygen to the cells, may have accounted for the symptoms

TABLE 4.3

Results from OECD POV and LRTP Screening Tool

Substance	Overall Persistence (days)	Characteristic Travel Distance (km)	Transport Efficiency (%)
ONCB	90.1	5,080	20.7
1,4-DCB	60.0	16,000	2.98
HCB	451	18,900	6.09
Benchmark	*195*	*5097*	*2.248*

that required workers to be hospitalized back in 1923. IARC found in 1999 that there is inadequate evidence in either experimental animals or humans for the carcinogenicity of chloronitrobenzenes [30]. Since then, concerns have been expressed about the potential carcinogenicity of the substance [27]. The US EPA has developed a cancer slope factor for ONCB [37].

Studies have shown moderate acute and chronic toxicity to aquatic life. A $PNEC_{aqua}$ has been derived as 0.026 mg/L [27].

4.1.6 Regulatory Status

In the United States, CAS number 88-73-3, Benzene, 1-chloro-2-nitro-, is listed on the Toxic Substances Control Act (TSCA) Non-Confidential Inventory with no Significant New Use Rule (SNUR) [38]. It has been designated as a High Production Volume (HPV) chemical [39]. In the European Union (EU), ONCB has been notified under REACH as an intermediate and as a chemical produced or imported at 100–1,000 tonnes per annum. The REACH notifications refer to use as an intermediate in closed processes, and to use as a laboratory reagent [40]. In China, ONCB is listed on the Inventory of Existing Chemical Substances Produced or Imported in China (IESC); it is also listed in the Catalog of Hazardous Chemicals [41].

4.1.7 Environmental Data

Unlike many of the substances discussed in this book, ONCB has primarily been used as a chemical intermediate to synthesize other substances rather than being produced for use as a pure substance or in chemical formulations. That limits the potential for environmental release to manufacturing facilities rather than dispersed consumer use, in contrast to a substance such as 1,4-dichlorobenzene used in products including toilet blocks and air fresheners.

A search of the scientific literature and governmental publications suggests that few researchers have studied the occurrence or distribution of ONCB in the environment. A few publications describe evidence of ONCB in the Mississippi River in the United States, as discussed further below. ONCB has also been detected in studies of surface water in Germany and China.

This case study began with an anecdote regarding the release of ONCB to the Mississippi River from a manufacturing facility in the late 1950s and the resulting contamination of a public water supply. The Water Quality Network subsequently reported in 1962 that surface water samples from the Mississippi River no longer contained ONCB [42]. Studies of conditions at landfills near the former manufacturing plant some 40 years later under the Comprehensive Environmental Response, Compensation, and Liability Act (CERCLA) detected nitrochlorobenzene in landfill leachate at 21 parts per million but not in nearby samples from the Mississippi River. Nitrochlorobenzene was not considered to be a "chemical of interest" with respect to soil or groundwater in the associated studies [43].

Publicly available data suggest that Mississippi River sediments in the reach from St. Louis/Sauget to Cape Girardeau may not have been extensively tested for ONCB. Studies of the river sediments near the former manufacturing site did not consider ONCB to be a "chemical of interest" [43]. The US Geological Survey has compiled data on sediment quality collected from 1974 through 2000 in the Upper Mississippi River in order "... to facilitate the assessment of riverine contamination, to increase the availability of historic sediment-contaminant data, to facilitate responses to contaminant issues in the Upper Mississippi River System, and to focus management and scientific efforts pertaining to contaminated-sediment." The database does not contain information on ONCB [44].

One study in the early 1980s detected ONCB in the tissues of fish collected between St. Louis/Sauget and Cape Girardeau. The concentrations reportedly ranged from 0.006 to 0.24 milligrams per kilogram (mg/kg), with the concentrations decreasing with distance from St. Louis/Sauget [45].

Studies of the environmental concentrations of ONCB in other parts of the world appear to be even more limited. Researchers reported in 1990 that they had detected ONCB in water samples from the Rhine River at Mainz, Germany at 0.08 to 0.12 micrograms per liter [46]. A subsequent study of surface waters in Germany between 1991 and 2000 showed that the environmental concentration of ONCB (90th percentile) was in the range of <0.005 to 0.58 micrograms per liter [47].

A study circa 2006, of water quality at monitoring stations on the Yellow River in China, detected ONCB in samples from 9 of 12 stations at concentrations ranging up to 2.518 micrograms per liter [48].

4.1.8 Synthesis

In the late 1950s, ONCB from an industrial discharge on the order of 20,000 pounds per day contaminated water in the Mississippi River. The residents of Cape Girardeau who used the Mississippi River for their water supply may have ingested ONCB at concentrations on the order of 30 ppb with their drinking water.

Regulators at that time had limited resources to assess or regulate the release. They recognized the possible hazard based on limited data suggesting that a lethal dose was between 5 and 50 mg/kg body weight. Subsequent studies have refined our understanding of the possible hazards. Based on the possible carcinogenicity of ONCB, the US EPA has published a nonenforceable limit on ONCB in drinking water of 0.23 ppb [49]. This guideline is two orders of magnitude below the concentration detected in drinking water in 1958.

It is remarkably difficult to piece together what happened since then. Surface water monitoring data showed the concentrations in the Mississippi River diminished after the manufacturer voluntarily ceased the discharge. Fugacity calculations suggest that the residual ONCB in the environment would have degraded with relative rapidity. Therefore perhaps the residuals from the former releases have disappeared. Long after the release in the 1950s, the US EPA regulated discharges of chlorinated benzenes to surface water under the Clean Water Act (40 CFR 401.15) and emissions to air from the manufacture of ONCB (40 CFR 401.15) under the Clean Air Act. The finding in the early 1980s that fish collected between St. Louis and Cape Girardeau contained ONCB in their tissues adds to the puzzle; the source of the ONCB in the fish tissue is unclear given the regulations on discharges and the expectation that historical releases would have biodegraded.

This case study illustrates above all the challenge of applying analytical life cycle thinking to the real world, particularly in hindsight. Quantitative production information, the foundation for any mass balance, is difficult to find. Although understandable from a business perspective, the paucity of information makes life cycle evaluations more difficult. More remarkable, perhaps, is the intermittent attention paid to ONCB in the Mississippi River under a patchwork of laws: through the predecessor to the Clean Water Act in 1959; the forerunner of TSCA in 1970; and, in a portion of the area of concern, under CERCLA at the former source from the 1990s through the present day.

TSCA does not impose limitations on the manufacture or import of ONCB. But market forces have made that nearly irrelevant in the United States. As of 2012, ONCB is no longer manufactured in the United States; a single company reportedly imported the substance [29]. ONCB is a chemical intermediate, and releases while manufacturing products made from ONCB are regulated under the Clean Water Act and the Clean Air Act. Thus the potential for uncontrolled release or exposure is now limited in the United States.

4.2 1,4-Dichlorobenzene

Like ONCB, 1,4-dichlorobenzene (1,4-DCB) first caught the attention of regulatory agencies early in the era of modern environmental directives. Unlike

ONCB, which has been primarily used as an intermediate, 1,4-DCB has been used in consumer products with consequent potential for human and environmental exposure. Such exposures ultimately led to restrictions on its use in the European Union. This case study compares how scientists and regulators viewed this substance in two eras, the mid-1970s and some 40 years later.

4.2.1 Introduction

One of the first reports issued by the newly formed US EPA, just before the passage of TSCA, described the use of 1,4-DCB circa 1975 [50, p. 324]:

> Para-dichlorobenzene in its solid form, > 99% purity, is used for moth control and space-deodorants and sublimes at 50 F to a heavier-than-air vapor. Action: fumigant against moths and their larvae, vapors act as a space deodorant, and a fungistat. As a moth control agent, a usage of 1 lb./100 cubic feet of confined space is recommended. ... The p-dichlorobenzene should be kept at a level such that the fumes in the confined area are concentrated enough to cause irritation to the eyes.

The report also describes the possible consequences of such use:

> Much of this chlorinated hydrocarbon is placed directly into wastewater via lavatory use and into the air we breathe at home and at work. Studies of the environmental aspects are negligible to date, and the conventional method used for analysis of chlorinated hydrocarbon pesticides in water does not normally detect dichlorobenzene. It has, however, been detected in the blood of workers exposed to it regularly and appears to accumulate in fatty tissues like other chlorinated hydrocarbons. (p. 354)

Chapters 2 and 3 of this book described some of the environmental and regulatory aspects of 1,4-DCB (CAS 104-46-7,* also known as para-dichlorobenzene or p-dichlorobenzene). The narrative that follows traces the issues that evolved from the 1975 report by the US EPA to the final decision by the European Commission in 2014 to ban certain uses of the substance.

4.2.2 Physicochemical Characteristics

Table 4.1 summarizes the key physicochemical properties of 1,4-DCB and the other chlorinated benzenes discussed in this chapter. Based on these data, 1,4-DCB is less soluble, more volatile, and more likely to bioaccumulate than ONCB.

* 1,4-dichlorobenzene may also be characterized as a chloro derivative of benzene (CAS 68411-45-0) or benzene, dichloro (CAS 25321-22-6).

4.2.3 Manufacture and Use

1,4-dichlorobenzene is essentially a byproduct of monochlorobenzene production. Manufacturers synthesize 1,4-DCB by reacting chlorine with benzene in the presence of a catalyst, usually ferric chloride. The reaction typically produces monochlorobenzene (approximately 85%), 1,4-DCB (approximately 10%), and 1,2-DCB (approximately 5%). The proportions of the isomers produced depend in part on the catalyst used and on the relative amount of the chlorine feed [51]. However [52, p. iv–30], "The highest volume chlorobenzenes (monochlorobenzene, o- and p-dichlorobenzenes) are produced as complementary products. While the proportions of the products can be varied fairly widely, none can go to zero. That is, if one product is desired, all must be produced, if only in minor quantities." In short, one of these substances cannot be produced without producing the others, which provides an economic incentive for finding uses for all of the reaction products.

Table 4.4 summarizes the available production data for DCB. The tabulated data do not make it explicitly clear, however, that the centers for manufacturing 1,4-DCB have shifted. Manufacture has ceased in the United States and diminished in the European Union, with chlorobenzene supplies increasingly coming from China and India [59,62,63]. We return to these production data and their environmental implications later in this chapter.

Manufacturers have found many uses for 1,4-DCB. In the mid-1970s, when this story starts, the US EPA reported that about 80–90% of the amount consumed in the United States went to "nonintermediate dispersive uses," that is, air fresheners (sometimes referred to as space deodorants) and moth control. (Those who remember American fashions of the 1970s will appreciate that the agency qualified this information by noting that "[u]se of 1,4-dichlorobenzene as a moth control agent for wool is declining as more synthetic fabrics are produced" [53, p. 67].) Deodorizing products included those for toilets and garbage and diaper pails. Approximately 10% of the 1,4-DCB was used to manufacture polyphenylene sulfide (PPS) resins. The compound was also used to control mildew, tobacco blue mold, bark beetles, moths and peach tree borers, lice, mites, and cockroaches. Other uses at the time included a disintegrating paste for molding concrete and stoneware, and as an intermediate in dye production [52, 53].

The usage of 1,4-DCB in the United States changed over time. By the late 1990s, authorities reported that although 1,4-DCB was used principally (25–55% of all uses) as a space deodorant for toilets and trash containers, and to control moths, molds, and mildews, other uses were increasing. The production of PPS resin accounted for approximately 50% of 1,4-DCB consumption, and the production of chemicals such as 1,2,4-trichlorobenzene another 10% [55]. As described below, the use of 1,4-DCB as a monomer in polymer production had continued to grow by 2013. As of 2014, a household products database listed only three products available in the United States containing 1,4-DCB, all moth repellants [64].

TABLE 4.4
Global Production of 1,4-DCB

	Approximate Annual Production (Metric Tonnes)				
Year	Global	United States	European Union	China	Reference
1962		33,800			[53]
1963		33,800			[53]
1964		28,700			[53]
1965		29,800			[53]
1966		30,100			[53, 54]
1967		30,200			[53, 54]
1968		31,900			[53, 54]
1969		23,600			[53, 54]
1970		31,600			[53, 54]
1971		31,900			[53, 54]
1972		35,100			[53]
1973		28,400			[53, 54]
1975		20,800			[53]
1976		16,600			[54]
1977		29,500			[53]
1978		18,700			[53]
1981		6,800			[55]
1983		28,300 Prod or import			[56]
1985			22,950		[57]
1986		34,000[a]			[58]
1987		35,000[a]	20,500		[57, 58]
1991			16,400		[57]
1993		32,600			[55]
1994			22,500– 30,500		[59]
1996				10,000	[60]
1998		23,000– 45,000			[55]
2002	301,000	23,000– 45,000			[55, 61]
2003	310,000				[61]
2007	37,000 est.				[61]
2010			30,000		[59]
2012		0[b]			[62]

[a] Demand.
[b] None produced; 23,000 to 45,000 tonnes imported.

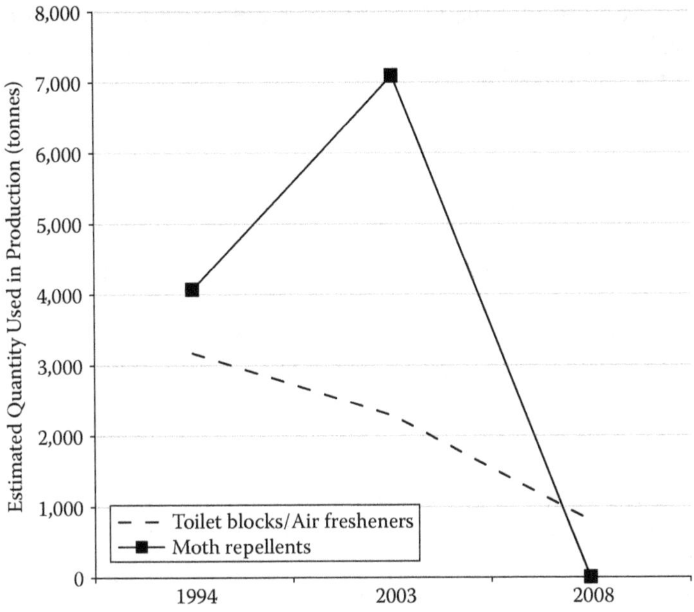

FIGURE 4.3
Trends in consumer uses of 1,4-DCB in the European Union.

Manufacturers in the European Union have utilized 1,4-DCB for many of the same purposes as users in the United States. Of the 14,494 tonnes per year consumed circa 1994, 49% was used as an intermediate, 22% toilet blocks/air fresheners, 28% moth repellents, and 0.7% grinding wheels. The intermediate use was primarily in the production of 1,4-dichloro-2-nitrobenzene, a precursor for dyes and pigments [57]. Consumer uses declined after 1994 as shown in Figure 4.3 [65]. Most of the mass produced in the European Union was exported for use in polymer production; some of the polymer was then reimported to the European Union [65].

Manufacturers in the United States, Japan, and China had increased their use of 1,4-DCB to produce PPS resins and other heat-resistant polymers. The resins are used in automotive and aircraft manufacture and in coal power plants for exhaust pipes and gas filter bags [59]. These burgeoning uses of 1,4-DCB bring new perspectives on life cycle effects and sustainability into this discussion of 1,4-DCB: replacing metal parts in automobiles and aircraft with lighter-weight polymer components can reduce fuel usage; the life cycle impacts of air pollution control equipment on coal-fired power plants are complicated indeed.

No data were identified during research for this book that quantified the usage of 1,4-DCB in China.

This rather dry recitation of data on the uses of 1,4-DCB illustrates two key points. First, usages can change over time, affecting the life cycle of a chemical. Second, it is difficult to obtain quantitative usage data to assess the life cycle of a chemical and the potential for exposure, even for a well-studied compound like 1,4-DCB.

4.2.4 Fate and Transport in the Environment

This section describes how our understanding of the behavior of 1,4-DCB in the environment has evolved over the 40-year history described in this case study. After an overview of the fate and transport of 1,4-DCB from a contemporary perspective follows a summary of early environmental mass balances that predated the practice of life cycle analysis, and a view of environmental fate and transport from the perspective of the EU restriction under REACH.

4.2.4.1 Overview

EpiSuite 4.11 predicts the environmental fate of 1,4-DCB pictured in Figures 4.1 and 4.2. Although these modeling results do not absolutely predict the fate and transport of 1,4-DCB, they do indicate its tendency to partition between compartments and biodegrade.

Based on the criteria used to characterize a chemical as persistent under the Stockholm Convention on Persistent Organic Pollutants [34], 1,4-DCB would not persist in water or soil but has the potential to persist in sediment and to be transported long distances by air. It would not be expected to bioaccumulate. The results from the *OECD Pov and LRTP Screening Tool* (Table 4.3) also indicate that the compound is not very persistent in air, water, or soil, but suggest that emissions could perhaps travel thousands of kilometers by air before completely degrading.

4.2.4.2 Materials Balance: 1979 and 1980

In a forerunner of the current practice of life cycle analysis, the US EPA commissioned a mass balance on chlorobenzenes in 1979. The study was intended "to determine, within the constraints of time and information availability, the quantities of chlorobenzenes annually released to the environment and the sources of those releases" [54, p. 1-1]. At that time, regulators knew relatively little about the environmental fate of 1,4-DCB, simply that it "[u]ndergoes degradation at a moderate to rapid rate. It is degradable by biological organisms, non-biological factors, and sunlight. ... There are no data as to persistence in soil" [50, p. 326].

The study report does not describe the methods in detail or document the calculations, but cautions the reader that it was relatively rudimentary and not meant to be a definitive study [54]. Figure 4.4 summarizes the material balance. Based upon the values shown in the figure, it appears that the material balance simply looked at releases based on usage, which were primarily to air and water, and did not assess transport between environmental

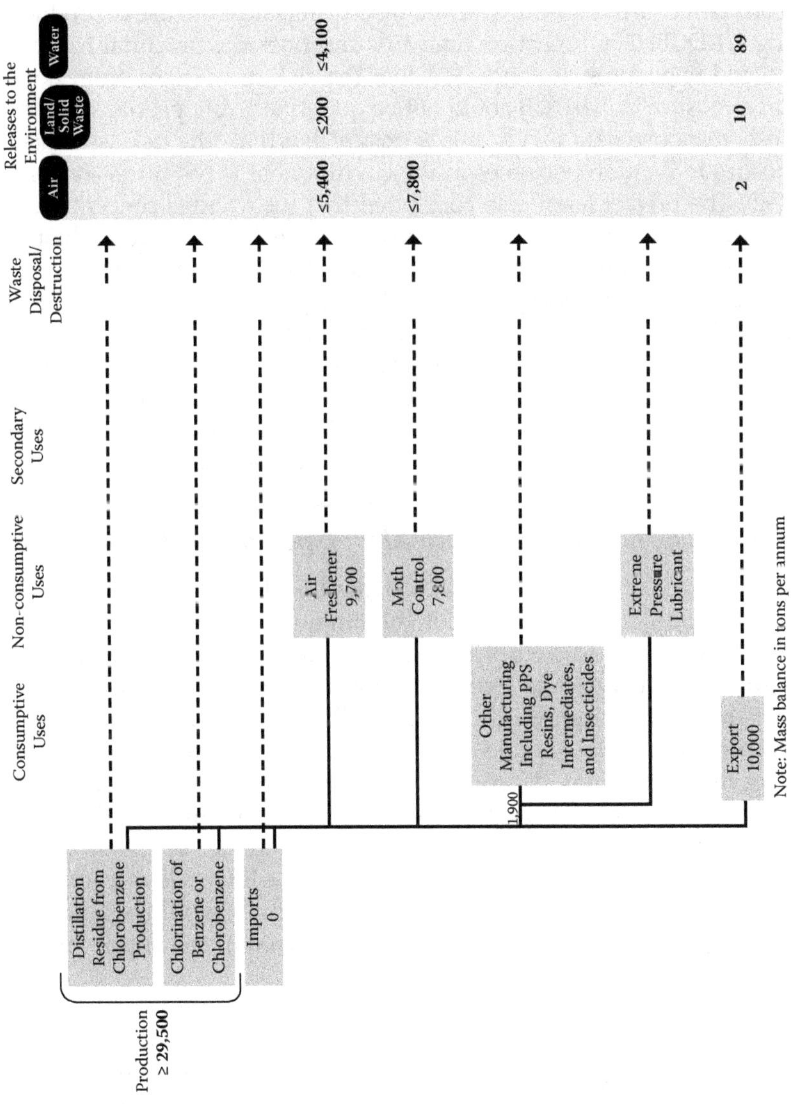

FIGURE 4.4
1979 Material balance on 1,4-DCB in the United States.

compartments or the potential for biodegradation to occur. By current terms, this material balance may seem relatively crude, but in 1979 it was a revolutionary look at our society's use of a chemical.

A year later, the US EPA did a more detailed mass balance and evaluated the consequences of exposure based on available monitoring data [66]. The mass balance considered emissions during manufacturing and from use of products containing 1,4-DCB. The report on the work describes the potential hazards and estimated exposures. It concluded that the risk to humans from environmental exposure to 1,4-DCB could not be quantified, due primarily to the lack of dose–response data for chronic exposure, but that "the risk associated with exposure to dichlorobenzenes in the environment is not unreasonable" (pp. 7-1–7-2). The project team also concluded that the average concentration of DCBs in the environment "would not pose a severe threat to aquatic biota," (p. 7-9) although fish kills had been reported as a result of direct discharges of large quantities of DCBs. Shortly after that, EPA did a companion study of the occurrence of DCB in drinking water, food, and air [67].

These early life cycle assessments provided estimates of exposure and risk in the United States that do not appear to have been revisited since. The US EPA has collected some relevant data under the Toxics Release Inventory (TRI) Program, however, which are presented in Section 4.2.7.

4.2.4.3 Global Supply Chains and Potential for Exposure: 2004–2010

In its early mass balance work on 1,4-DCB, the US EPA acknowledged that the substance could be imported (but the amounts were negligible) and some of the material produced in the United States was exported. But the analyses focused on uses and emissions within the United States. When the European Commission examined 1,4-DCB under REACH, the supply chains for products containing 1,4-DCB reflected the globalization of business during the intervening decades. Figure 4.5 shows the supply chain for deodorizer products sold in the European Union circa 2010 [59]. Defining the system boundaries on an environmental mass balance of 1,4-DCB had become much more challenging. An evaluation of the life cycle impacts of a substance in one geography, such as the European Union, would not include the environmental releases or effects from other steps in the supply chain that occurred elsewhere.

The EU Risk Assessment included an estimate of emissions of 1,4-DCB during production and use *within the European Union* during the early 1990s as shown in Table 4.5 [57]. (Recall from the previous discussion that of the 14,494 tonnes of 1,4-DCB consumed per year at that time, 49% was used as an intermediate, 22% toilet blocks/air fresheners, 28% moth repellents, and 0.7% grinding wheels.) Although the manufacture of air fresheners and moth repellants accounted for about half of the consumption of 1,4-DCB, the manufacture and use of those products accounted for just over 90% of the estimated releases.

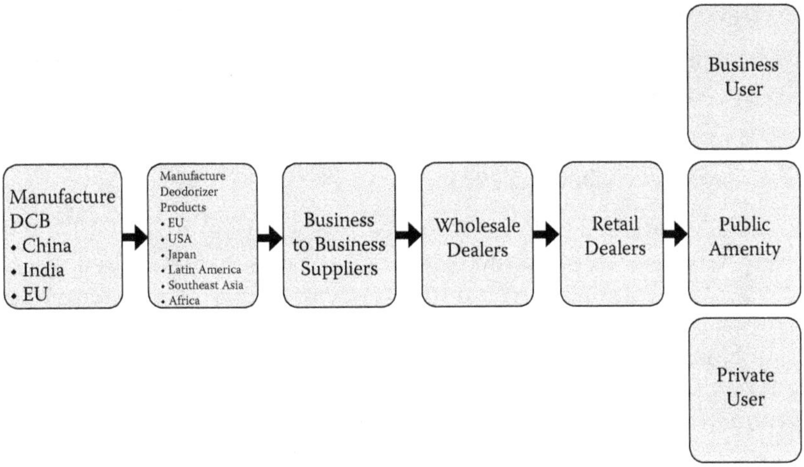

FIGURE 4.5
Supply chain for deodorizer products containing 1,4-DCB.

TABLE 4.5
Estimated Releases of 1,4-DCB in the European Union

Source	Releases (tonnes/year)		
	Surface Water	**Atmosphere**	**Soil**
From production	3.6	50.5	Negligible
From use as an intermediate	3.45	2.3 Direct 2 via STP*	Negligible
From formulation of air fresheners or moth repellants	Negligible	72.4	Negligible
From use of air fresheners or moth repellants	420	6821 Direct 210 via STP	Negligible
From use in production grinding wheels	Negligible	100	Negligible

* STP: Sewage treatment plant.

The risk assessors used the EUSES model to estimate exposure concentrations from these emissions. At steady state, the model predicted the following environmental concentrations (PECs) [57]:

- $PEC_{regionalsurface\ water} = 0.04\ \mu g/l$
- $PEC_{regionalsediment} = 0.4\ \mu g/kg$ wet weight $= 1.0\ \mu g/kg$ dry weight (dw)
- $PEC_{regionalair} = 0.074\ \mu g/m^3$
- $PEC_{regionalagricultural\ soil} = 0.008\ \mu g/kg$ wet weight $= 0.01\ \mu g/kg$ dw

Based on a comparison to measured concentrations, the assessment team concluded that the estimated concentrations were reasonable [57].

4.2.5 Overview of (Eco)Toxicological Concerns

The understanding of the (eco)toxicity of 1,4 DCB evolved markedly between the 1970s and the present time, as described below.

4.2.5.1 Early Understanding: 1975

The US EPA compiled data on the effects of exposure to 1,4-DCB in 1975 [50]. 1,4-DCB was considered to be slightly toxic to mammals by ingestion. Exposure by direct contact would irritate the eyes, mucous membranes, and skin, albeit "only at levels not ordinarily voluntarily tolerated by humans" (p. 326). No data were available regarding toxicity to fish, wild mammals, or soil organisms, although 1,4-DCB was known to be relatively nontoxic to lower aquatic organisms and moderately toxic to birds. It did not bioaccumulate. The agency concluded its evaluation of 1,4-DCB by stating "when used as specified, it provides relatively safe and inexpensive moth protection and space deodorant properties."

4.2.5.2 EU Risk Assessment and Classification under REACH: 2004–2009

1,4-DCB underwent a risk assessment in the European Union which was published in 2004 [57]. As a substance produced or imported into the European Union at 10,000–100,000 tonnes per annum, 1,4-DCB was also notified under REACH with a full suite of data in 2009 and the hazards characterized under the regulation on Classification, Labelling, and Packaging of substances and mixtures (CLP). These efforts assembled vastly more data on the ecotoxicity and mammalian toxicity of 1,4-DCB than were available to the US EPA in 1975.

The European Union classified 1,4-DCB as follows [65]:

- Category 2 carcinogen
- Eye irritation, hazard category 2
- Hazardous to the aquatic environment, acute hazard category 1
- Hazardous to the aquatic environment, chronic hazard category 1

This characterization of hazards differed markedly from the US EPA's 1975 precis, which did not discuss carcinogenicity and reflected limited aquatic toxicity data. Regulators working under REACH had a far different technical basis for decision making than had the US EPA in the early days of TSCA.

4.2.6 Regulatory Status

An existing chemical when TSCA was passed, 1,4-DCB came in for early scrutiny under the law. TSCA established the TSCA Interagency Testing

Committee to identify and recommend to the US EPA Administrator which substances should be tested to determine their hazards to human health and the environment. The committee's 1977 report recommended that mono- and di-chlorobenzenes be tested for carcinogenicity, mutagenicity, teratogenicity, other chronic effects, and environmental effects; in addition, the Committee recommended an epidemiological study. The committee based this recommendation on the relative paucity of data for 1,4-DCB, which they regarded as serious in light of an estimated environmental release rate of up to 71,000,000 pounds and exposure by over a half million workers [68]. After assessing the potential economic impacts of requiring testing [52], the agency proposed that 1,4-DCB be tested under Section 4 of TSCA. The US EPA subsequently proposed a testing rule for public comment. In 1986, perhaps recognizing other testing efforts, the US EPA promulgated rules requiring only soil adsorption coefficient testing and reproductive effects testing [57,69]. In 2004, the US EPA promulgated a second rule requiring in vitro dermal absorption toxicity testing [70]. Although the agency has posted the results [71], it has not assessed risks based on those results [72] or taken any further regulatory action under TSCA. (Under the Federal Insecticide, Fungicide and Rodenticide Act [FIFRA], however, the US EPA has assessed the risks from exposure to 1,4-DCB when used as a pesticide. This assessment is described further below.)

The 1998 Chemical Right to Know Initiative launched the Voluntary Children's Chemical Evaluation Program (VCCEP) in 2000. Under this program the US EPA asked manufacturers and importers to provide information voluntarily on 23 chemicals, including 1,4-DCB, to which children were expected to have a high likelihood of exposure. The requested data included information on health effects, exposure, risk, and data needs. In 2001 an industry consortium agreed to evaluate available existing hazard and exposure data (Tier I of the VCCEP) and to develop data if necessary. As of 2014, no VCCEP reports on 1,4-DCB appear to have been published [73,74]. The Inspector General of the US EPA has criticized the program for not producing complete or timely results [74]. To be fair, the need for results under the VCCEP may have decreased as consumer uses of 1,4-DCB and the resulting potential for exposure have declined.

1,4-DCB is registered as a pesticide under FIFRA for uses such as in moth repellants to protect garments from insect damage and for the control of lice and ticks in and around birdcages. First registered in the United States in 1948, the substance was reregistered in 2008 after an assessment of the potential risks to human health resulting from these uses. The registration specifies certain labeling and handling precautions for pesticidal products containing 1,4-DCB [75].

In the European Union, authorities first considered the need to regulate 1,4-DCB under the Dangerous Substances Directive and assessed the risks from exposure to 1,4-DCB under the Existing Substances regulation, beginning in the mid-1990s. In a decades-long process interrupted by the passage of

REACH, authorities determined that certain uses should be restricted based on the risk characterization described in Section 4.2.5.2. REACH Annex XVII was amended in January 2014 to include the provision that 1,4-DCB "[s]hall not be placed on the market or used, as a substance or as a constituent of mixtures in a concentration equal to or greater than 1% by weight, where the substance or the mixture is placed on the market for use or used as an air freshener or deodoriser in toilets, homes, offices or other indoor public areas." (For additional discussion, see Chapter 3.)

1,4-DCB in mothballs is considered a biocide and is regulated as such. The authorities in Spain restricted the use of 1,4-DCB in biocides in 2005, and subsequently the substance was not registered in Annex I, IA, or IB of the Biocidal Products Directive. Consequently 1,4-DCB cannot be used now in mothballs in the European Union [59].

In contrast to the lengthy regulatory history in the United States and the European Union, the story of 1,4-DCB under chemical control regulations in China is quite brief. 1,4-DCB is listed as an existing chemical substance in China. It is also considered to be a hazardous chemical [76].

4.2.7 Environmental Data

Data from the TRI program in the United States offer a contemporary counterpoint to the mass balance calculations performed in 1979 and 1980 and described above. The TRI program requires many industries to report the amounts of certain chemicals that they release to the environment and/or manage through recycling, energy recovery, and treatment each year. The data have limitations. Not all industries must report releases, and the data do not reflect releases that occur from consumer uses. Further, the US EPA cautions that the TRI data do not indicate whether (or to what degree) the public has been exposed to those chemicals. Figure 4.6 summarizes the TRI data for 1,4-DCB [77]. The identity of the reporting companies and the North American Industry Classification System (NAICS) Code which each company chooses offers some clues to the changes in the marketplace. In 1989, 11 companies reported releasing more than 5,000 pounds per year of 1,4-DCB. Four manufactured chemicals, four abrasive products or grinding wheels, and three sanitation goods or soaps. By 2012 only two companies reported releases over that threshold. One manufactured chemicals and the other, plastics and resins. The dramatic decrease in emissions between 1989 and 1991 shown in the graph results from two plants, one a chemicals manufacturer and the other a manufacturer of abrasive products/grinding wheels, each of which reduced emissions by over 98%. The reason for the reduction cannot be discerned from the TRI data.

Between 1990 and 2005, the US EPA measured the concentration of 1,4-DCB at ambient air monitoring stations around the country. These data showed a decrease in the levels of 1,4-DCB, particularly from 2000 to 2005 when the median annual concentration decrease was 9%. The distribution trends

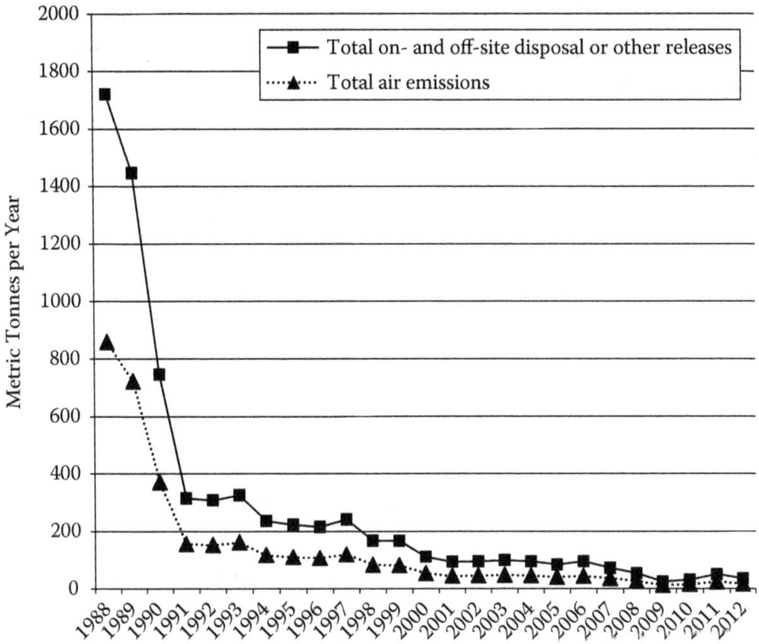

FIGURE 4.6
Industrial releases of 1,4-DCB in the United States.

across monitoring sites varied greatly, perhaps due to analytical uncertainty because the concentrations detected were near the method detection limit [78]. Nonetheless, the decreasing concentrations of 1,4-DCB in ambient air appear to be consistent with decreasing industrial emissions and consumer uses of products containing 1,4-DCB over that time.

The output of the *OECD Pov and LRTP Screening Tool* (Table 4.3) suggests that although much of the emitted 1,4-DCB would degrade, roughly 3% of the released mass of 1,4-DCB may have been transported very long distances and deposited in remote regions. As shown in Figure 2.4, atmospheric transport processes might carry any emissions from the United States that did not degrade to the Arctic. Once transported there, 1,4-DCB would likely degrade more slowly than predicted by models such as the *OECD Pov and LRTP Screening Tool.* Most calculations rely on default data regarding degradation rates at temperatures of approximately 20 to 25°C. The colder temperatures in the Arctic would slow degradation relative to such estimates.

Few studies have examined the concentrations of 1,4-DCB in the northern environment. A literature search identified one report which demonstrated that 1,4-DCB can partition from air to ice and snow based on laboratory experiments, but no field data were reported [79]. The Fish Monitoring Program of the Alaska Department of Environmental Conservation analyzed samples of 157 fish (caught in southeast Alaska, Gulf of Alaska, Bering Sea, and

Kotzebue) for 1,4-DCB, among other substances. Chemists detected 1,4-DCB in 121 samples. The Department of Environmental Conservation concluded that these data represented a global presence of 1,4-DCB. Upon reviewing the data from the fish biomonitoring project, the Alaska Division of Public Health, Section of Epidemiology, concluded that the low levels detected did not present a public health concern [80]. Although intriguing, these data are too limited to draw conclusions about the possible transport of 1,4-DCB to the far North and its persistence there.

4.2.8 Synthesis

> Agents who operate in complex environments must often construct plans on the basis of incomplete knowledge. In such situations, the successful agent must incorporate into his plans actions which obtain information. These plans are intrinsically sketchy to begin with and become more specified as the agent proceeds through his plan. (p. 867)

Although written within the context of developing artificial intelligence [81], this passage neatly sums up the progression illustrated by this case study. The US EPA singled out 1,4-DCB soon after passage of TSCA in the 1970s to evaluate releases to the environment and identified the need for testing. Those early analyses did not lead to regulation of the substance to restrict exposure, at a time when the predominant uses of 1,4-DCB, in odor control products and mothballs, posed the potential for consumer exposure and released the substance to air and wastewater.

Decades later, regulators in the European Union began to explore the risks from consumer exposures more closely. By that time far more data were available to quantify the possible risks to human health and the environment, including concern over the potential carcinogenicity of the substance. Over a process that lasted some 20 years, the European Union concluded that certain consumer uses should be banned.

During that time, consumer preferences had already shifted the usage of 1,4-DCB away from uses for moth protection and deodorants. As one authority wrote [59, p. 23], "...products based on 1,4 dichlorobenzene are considered to be 'old fashioned' and perhaps more acceptable to older generations which are used to the strong, moth ball-like odour of the substance." Further, the recognition that 1,4-DCB might be a carcinogen may have deterred some users [59]. Consequently by the time regulations took effect they would have little real impact on consumer uses.

More and more 1,4-DCB has been used to produce lightweight plastics that could replace metal parts in automobiles and airplanes, with the consequent reduction of petroleum usage. Use as a chemical intermediate to synthesize other products generally releases a far lower proportion of 1,4-DCB to the environment than other uses, based on the common-sense recognition that deodorant and moth-protection applications relied on the product subliming and based on the data shown in Table 4.5.

What of the long-term environmental consequences? As described above, 1,4-DCB would not be expected to persist in water or soil but has the potential to persist in sediment and to be transported long distances by air. Consider the following thought experiment. Presume that a country has produced 1,4-DCB at the quantities shown in Table 4.4 for the United States. (For years when no data were available, they were interpolated from available data for this thought experiment.) Assume that the amount of 1,4-DCB used in air fresheners and moth repellants comprised 90% of the total annual production in the 1960s, 80% in the 1970s, 70% in the 1980s, 50% in the 1990s, and 20% since. Estimate the amount released to the environment from this use as 97% of the mass, consistent with the EU risk assessment (Table 4.5). Finally, distribute the mass between environmental compartments and assume biodegradation occurs according to the projections of the EpiWin model (Figure 4.1). Figure 4.7 shows the results of this mass balance.

Although the calculations are far too crude to represent a real-world situation (notably neglecting the time required to reach steady state) this thought

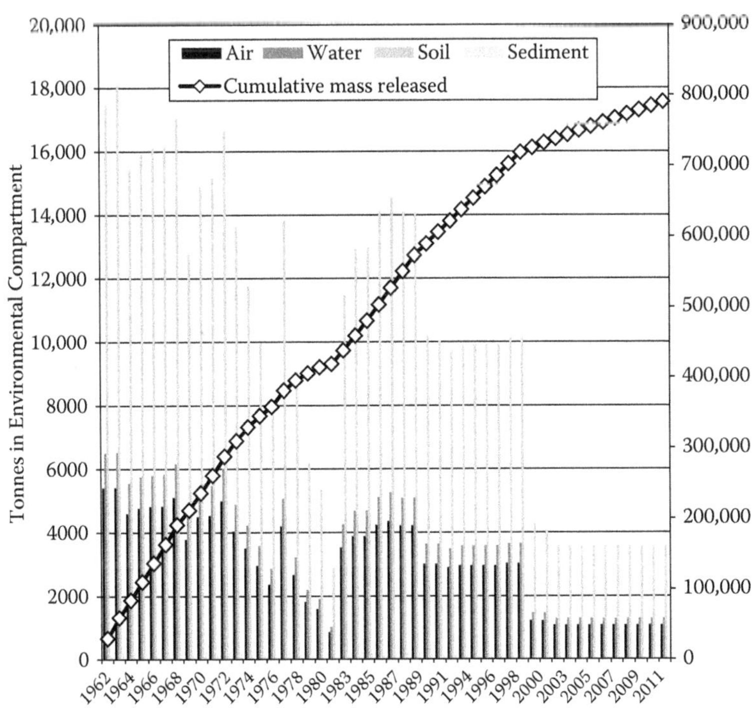

FIGURE 4.7
Thought experiment: hypothetical release, distribution, and degradation of 1,4-DCB.

experiment illustrates several points. The first point is perhaps not intuitively obvious. Environmental pollutants are not always where we might expect to find them. Although deodorizers and mothballs release 1,4-DCB to the environment primarily by volatilization, at steady state this compound partitions to soil. Second, because the compound biodegrades fairly rapidly the mass in each environmental compartment is highly sensitive to the amount of ongoing discharge. Even though the total tonnage released appears substantial, only a fraction lingers in the environment according to these simple calculations. Reducing the emissions of a compound that is not persistent or bioaccumulative can result in a relatively rapid decrease of the amount in the environment. The wild card in this thought experiment, of course, is the recognition that this simple calculation neglects global transport by air and the decrease in the rate of degradation that would be expected at colder temperatures.

To put the information on 1,4-DCB into further context, let's look at a more highly chlorinated substance.

4.3 Hexachlorobenzene

This case study reviews the history of hexachlorobenzene from its heyday as an agricultural boon to its global ban a few decades later as one of the "dirty dozen" chemicals.

4.3.1 Introduction

Wheat bunt can decimate a crop. Some farmers call this fungus "stinking smut" for its ability to reduce healthy grain to a fishy-smelling black powder. The plague can persist in contaminated seed and may contaminate soil to ruin the next year's harvest. Farmers watched the fungus destroy crops for hundreds of years with no relief until mercury-based pesticides provided a partially effective treatment. Then in the middle of the last century the synthetic chemical industry produced a new and highly effective solution: hexachlorobenzene (HCB; CAS number 118-74-1) [82, 83, 84].

4.3.2 Physicochemical Characteristics

Table 4.1 summarizes essential information about critical properties of HCB and Figures 4.1 and 4.2 present the output of the EpiSuite 4.11 model. With its higher degree of chlorination, HCB is less soluble and more persistent than ONCB or 1,4-DCB, and can partition into fat and bioaccumulate to a markedly greater degree.

4.3.3 Manufacture and Use

A scientist writing in 1945 about his work with a powerful new chemical could barely restrain his excitement, even in the dry prose of a scientific journal [85, p. 393]:

> Following closely on the several recent demonstrations of the considerable insecticidal potency of `D.D.T.' (dichloro-diphenyl-trichlorethane) it is surprising to find another substance which, on preliminary trial, promises to be at least as good, and in some ways even better than, `D.D.T.' ... I have recently had the opportunity of carrying out a few trials with a substance temporarily designated `666' which has been recognized for some little time by ... research workers as possessing unusually active insecticidal properties. This substance, which may now openly be referred to as hexachlorobenzene, has the general formula C6H6Cl6 and is a very remarkable compound in [its] extraordinary insecticidal properties ...

The use of this very remarkable compound to control wheat smut soon followed. Farmers also treated the seed of sorghum, onions, and other crops with HCB. The commercial fungicide formulations they used to do so typically contained 10–40% HCB. HCB also found various other uses, including as a component in pyrotechnics manufacture, synthetic rubber production, graphite electrode production, and the manufacture of other chemicals, including pentachlorophenol [85–87].

Chemical companies made HCB as a byproduct of manufacturing other chlorinated hydrocarbons such as carbon tetrachloride, tetrachloroethylene, and trichloroethylene. In the 1970s, for example, these processes accounted for 89% of the HCB manufactured in the United States [85]. HCB can also be produced directly by treating isomers of hexachlorocyclohexane with sulfuryl chloride or by reacting benzene or chlorobenzenes with chlorine [85].

Little information is available on historic production rates. Table 4.6 summarizes the available data. Production apparently peaked in the late 1970s or early 1980s. The use of HCB as a pesticide declined after that. The United States ceased large-scale production by 1984; the European Community in 1981; and China in 2004 [86,88,89].

HCB was also an impurity in certain other pesticides [86,90]. Dimethyltetrachloroterephthalate (DCPA, also known as Dacthal) contained up to 10% HCB and pentachloronitrobenzene (PCNB), 3%. Chlorothalonil, picloram, and pentachlorophenol reportedly contained lesser amounts of HCB. These pesticides were heavily used in the late 1970s and early 1980s; in the United States, for example, thousands of metric tonnes of these pesticides were applied every year. As a result, although the relative proportion of HCB in these products was relatively low, tonnes of HCB were likely released to farmland, gardens, lawns, and golf courses.

TABLE 4.6

Global Production of HCB

	Approximate Annual Production (Metric Tonnes)					
Year	Global	United States, as Pesticide	United States, Total	European Union	China[a]	Reference
1958		345				[85]
1959		327				[85]
1960		200				[85]
1973		318				[85]
1974				4,000		[88]
1975			1,452			[86]
1976				2,600		[88]
1977			2,700			[54]
1978	9,072			7,258		[88]
Early 1980s		<23				[86]
1988			0		5,443	[89]

[a] Production reportedly began in China in 1958 [89].

4.3.4 Fate and Transport in the Environment

HCB tends to sorb to soil rather than remain in water or air (Figure 4.2) but has a relatively high vapor pressure. It degrades slowly. Consequently HCB released to the environment will persist and a large amount of the mass may be transported long distances by air. (See Table 4.3 for relevant model results.) Remember the "grasshopper effect" described in Chapter 2? This effect accounts for the airborne movement of HCB from the countries where it was manufactured and used toward the poles of the earth [91,92].

Once transported to colder regions, HCB will degrade even more slowly. Figure 4.8 shows the half-lives of HCB in various media in temperate zones and the effect of temperature on the half-life in air [20]. As those data illustrate, contaminants such as HCB may persist once they have reached the Arctic or Antarctic for far longer than expected based on models calibrated to temperate conditions.

4.3.5 Overview of (Eco)Toxicological Concerns

When HCB emerged as an effective solution for wheat bunt, scientists obviously recognized its fungicidal and insecticidal properties but had not characterized the hazards to human health or ecosystems. One scientist, who published his results in an Italian journal in 1954, took an early look at the effects of exposure on mammals [93]. But the data from his study did not allow anyone to predict or prevent a tragedy that was about to unfold some 2,000 kilometers away.

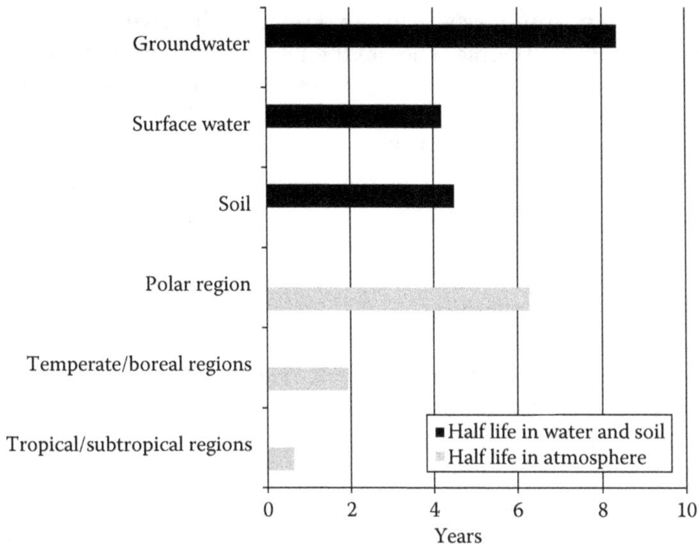

FIGURE 4.8
Half-lives of HCB.

This lack of data on the hazards of exposure was characteristic of the time; the US government, for example, did not require acute or subacute toxicity testing on pesticides until 1954 and did not require tests of carcinogenicity or reproductive effects until 1960 [94]. (Further, as described in Chapter 2, the use of a chemical such as HCB for nonpesticidal purposes was unregulated until decades later.) The potential hazards from exposure to HCB began to emerge in the late 1950s. The first of the vignettes related below was a tragic accident; the second resulted from waste disposal that may appear irresponsible by current standards but did not violate the laws of the time. Neither reflected the intended uses of HCB.

The government of Turkey purchased HCB-treated wheat for seed in 1954 [95]. That same year the wheat crop in Turkey failed [96]. When the treated seed arrived too late in the season for planting it was reportedly diverted for food. Each of the people eating bread made from treated wheat unknowingly ingested an estimated 0.05 to 0.2 grams of HCB per day [97,98]. Within six months of eating bread made from HCB-treated seed people developed porphyria cutanea tarda (PCT). Eventually an estimated 3,000 to 5,000 people, most of them children, developed PCT. Approximately 10% died. Stillbirths to exposed mothers were common and breastfeeding infants of exposed mothers often showed symptoms of a disease called "pink sore" or pembe yara; the mortality rate among these children was approximately 95% [95,97].

PCT is one of a class of diseases that cause a defect in the enzyme system responsible for heme production. (Heme is a component of hemoglobin, which carries oxygen in the blood.) The defect leads to the buildup of chemicals called porphyrins which can damage skin tissues. This skin damage manifested in the people who ate the HCB-contaminated wheat as photosensitivity, ulcerated skin, hyperpigmentation, and hirsutism. Their bones deteriorated [95,99]. Pictures of children with PCT [97] are hard to look at, even at a distance of more than 50 years; they show blistered and deeply scarred skin, blinded eyes, and twisted hands. The effects of exposure could be so extreme that some of the children were referred to as having "monkey disease" because of the degree of black discoloration and hair growth on their skin [95,99]. The acute outbreak of PCT subsided when the Turkish government ceased the distribution of HCB-treated grain in 1959. From our perspective, it may seem hard to believe that HCB-treated grain was distributed for so many years despite the incidence of PCT. But scientists did not realize the link between eating HCB-contaminated bread and PCT until 1958, or demonstrate the effect in laboratory animals until 1961 [100].

Exposure to HCB had long-term effects. Acute effects on skin ceased within a month after stopping exposure to HCB although scarring remained. Three to five years after exposure, over half the children studied had "'rheumatoid' arthritic change" [95]. Twenty-five to thirty years later, when doctors examined approximately 250 of the survivors, they found that some of the patients still showed evidence of disturbed porphyrin metabolism and lingering sensory motor damage. The milk of mothers in this group contained elevated levels of HCB [98]. Exposure to HCB may also have caused reproductive effects: a later evaluation of the proportion of male births in Turkey found some evidence that mothers who had developed PCT from exposure to HCB gave birth to a lower proportion of boys [101]. This incident in Turkey led scientists to understand the effects of human exposure to relatively high doses of HCB.

The potential for bioaccumulation of HCB and evidence that environmental contamination could be more significant than previously realized came to light in an incident in the United States in 1972. In December 1972 inspectors from the US Department of Agriculture (USDA) took routine samples of fat from beef carcasses in Louisiana for pesticide analysis. They found 1.5 parts per million (ppm) HCB. At that time, the USDA considered 0.3 ppm HCB in beef fat to be acceptable, apparently by analogy to dieldrin. The inspectors condemned those carcasses and broadened their investigation to testing hundreds of cows. Fat samples from 29% of the hundreds of cattle tested contained HCB above 0.5 ppm; the maximum concentration detected was 6.0 ppm. Hundreds of cows were destroyed and tens of thousands of cows were quarantined. The US EPA hypothesized that the cattle were contaminated via air emissions of HCB from a nearby landfill, and noted that the cattle with the highest concentrations of HCB in their fat had grazed alongside a roadway used to transport HCB waste from a manufacturing facility to the landfill. A limited sampling program indicated that soil near the landfill contained 5,000 ppm HCB and air samples approximately 14 ppb [87].

In contrast to these events and probably reflecting its focus on the intended use of HCB, the USDA Cooperative Extension Service advised farmers in the

United States as late as 1996 that "hexachlorobenzene is a practically nontoxic compound" [102]. A summary of the (eco)toxicity concerns regarding HCB as they are currently known follows.

Acute exposure to very high levels of HCB may affect the nervous system, potentially causing weakness, tremors, or convulsions; skin sores, and/or liver and thyroid effects in the people exposed. Chronic exposure, depending on the dose, can damage the liver and reproductive system and can cause developmental effects. HCB tends to bioaccumulate in fatty tissue [99]. The International Agency for Research on Cancer (IARC) has concluded that there is inadequate evidence in humans for the carcinogenicity of HCB but sufficient evidence in experimental animals for the carcinogenicity of HCB, and has classified HCB as possibly carcinogenic to humans (Group 2B) [103].

Many aquatic toxicity tests have found that the acute or chronic effect being tested did not occur at the maximum concentration of HCB that can be dissolved in water. The predicted no effect concentration (PNEC) in water, considering the effects of exposure on aquatic plants, invertebrates, and fish, is 0.37 micrograms per liter [104].

Toxicity testing of soil—where HCB is more likely to be found than in water—is less common. As described in Chapters 2 and 3, terrestrial toxicity testing is often not a part of formal risk characterization. Some scientists have tested the effect of exposure to HCB-contaminated soil on earthworms and found that HCB can bioaccumulate (see, e.g., [105]).

4.3.6 Regulatory Status

HCB differs from ONCB and 1,4-DCB when it comes to chemical control regulations: Unlike its less-chlorinated cousins, HCB has been singled out for restriction in international environmental treaties. International efforts to regulate pollutants such as HCB began in 1979 with the Convention on Long-Range Transboundary Air Pollution, often abbreviated as CLRTAP. Catalyzed by concerns over the long-range transport of sulfur compounds, the convention stated that [106, Article 2]: "The Contracting Parties, taking due account of the facts and problems involved, are determined to protect man and his environment against air pollution and shall endeavour to limit and, as far as possible, gradually reduce and prevent air pollution including long-range transboundary air pollution." CLRTAP declared the intent to research and share information on pollutants that could be transported long distances by air. Eight protocols have formally extended the scope of CLPTRP, one of them to include HCB.

The 1998 Protocol on Persistent Organic Pollutants (POPs), also known as the Aarhus Protocol after the city where it was created, stated the ultimate objective to eliminate any discharges, emissions, and losses of POPs. It specifically declared the intent to eliminate the production and use of HCB in the future [107]. The European Union signed the Aarhus Protocol in 1998 and approved it in 2004; the United States signed the protocol in 1998 but has never ratified it; China has not signed the protocol [108].

A second treaty on POPs broadened the focus beyond air pollutants that could migrate long distances [109]:

> Being aware that ... POPs pose major and increasing threats to human health and the environment, in May 1995 the Governing Council of [United Nations Environmental Programme] UNEP requested ... that an international assessment process be undertaken of an initial list of 12 POPs and that the Intergovernmental Forum on Chemical Safety (IFCS) develop recommendations on international action for consideration by UNEP Governing Council and World Health Assembly no later than 1997. In June 1996, IFCS concluded that available information was sufficient to demonstrate that international action, including a global legally binding instrument, was required to minimize the risks from the 12 POPs through measures to reduce and/or eliminate their emissions or discharges.

Negotiations that began as a result of this UNEP request ultimately led to the adoption of the Stockholm Convention on Persistent Organic Pollutants (POPs) in 2001. There have been 159 parties that have signed the Stockholm Convention. The European Union and China signed the convention in 2001 and ratified it in 2004; the United States also signed the convention but has not ratified it [110]. Parties to the Stockholm Convention agreed to, among other actions [111],

- Prohibit and/or take the legal and administrative measures necessary to eliminate the production, use, import, and export of the POPs specified in the treaty
- Take measures to reduce or eliminate releases of POPs from unintentional production
- Take measures to reduce or eliminate releases of POPs from stockpiles and wastes

The "dirty dozen" POPs initially listed in 1995 and restricted in 2001 included HCB. The Stockholm Convention (or more precisely, the law that each signatory must pass to enforce the Stockholm Convention) bans the production of HCB but allows its use as an intermediate and a solvent for pesticides. It also recognizes that many processes, ranging from certain kinds of manufacturing to the incineration of municipal waste, can unintentionally produce HCB [111]. While global treaties banning POPs were evolving at the United Nations between 1979 and 2004, nation-states took steps to regulate HCB. Those regulatory schemes reflected the dual use of HCB as a pesticide and as a chemical intermediate.

In the United States HCB was first registered as a pesticide under FIFRA and then listed as an existing substance under TSCA. The TSCA Inventory does not indicate any SNUR on HCB [38]. After contamination of cows in

Louisiana and similar concerns over the contamination of sheep in California, supplies of Dacthal containing 10% HCB were voluntarily taken off the market in California in the 1970s [112]. The manufacturers of registered HCB pesticides voluntarily requested cancellation of the nationwide registration of their pesticides in 1984 and 1985. The existing inventories of HCB pesticides could be used until exhausted or until July 1985, whichever came first [86].

Across the ocean, the use of HCB as a pesticide was banned under Council Directive 79/117/EEC of 21 December 1978, which prohibited the marketing and use of HCB in the European Union after 1981 but allowed its production for purposes of research or analysis or for export [113]. More generally, in 1997 HCB became regulated under the Limitations Directive which identified it as a carcinogen [114]. In 2004, Regulation (EC) No 850/2004 [115] solidified the European Union's commitment to the Stockholm Convention and its restrictions on HCB. HCB cannot now be placed on the market as a substance or mixture intended for consumer use (REACH Annex XVII, Entry 28 - Appendix 2 [116]).

In China, HCB is on the IECSC and the Catalog of Hazardous Chemicals [41]. China ceased production of HCB in 2004, according to its National Implementation Plan for Stockholm Convention on Persistent Organic Pollutants [117].

Readers paying careful attention to the dates of these various regulatory actions may have noticed that some countries had already banned broad usage of HCB by the time the Stockholm Convention was ratified. Then what was the point of including HCB in the Stockholm Convention? Or put another way, has the Stockholm Convention had any meaningful effect on the production or trade of HCB? Data on the global trade of HCB offer some perspective.

The United Nations tracks imports and exports of HCB and DDT, both banned under the Stockholm Convention, under the single commodity code 290362. The data are not easily aggregated on a global basis by a casual user of the UN's Commodity Trade Statistics Database, but it's a relatively simple matter to download information on the exports from a single country. Figure 4.9 shows the annual exports under commodity code 290362 from selected countries [118]. The data show a general decrease in exports of HCB/DDT from the United States and the European Union and appear to show an increase in the exports from India and China except perhaps for the last few years reported. (The exports from one country in one year were so high by comparison to all the other values that the export data had to be plotted on logarithmic scale: in 2008, India reportedly exported 39,200 tonnes of HCB/DDT, 82% of which went to the United Arab Emirates. India ratified the Stockholm Convention in 2006, and the United Arab Emirates, in 2002.) The only clear conclusion one can draw from the data shown in Figure 4.9 is that the strictures in the Aarhus Protocol and the Stockholm Convention on the production, import, and export of the "dirty dozen" have not stopped the shipment of HCB/DDT around the world. Environmental data, as described below, provide another perspective on the effect of regulatory bans on HCB.

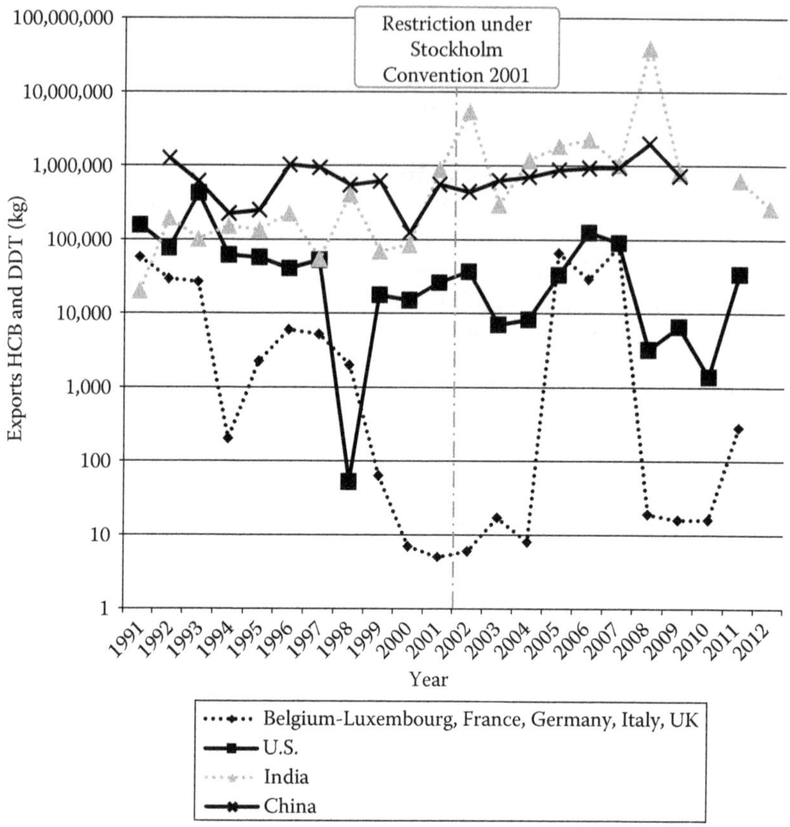

FIGURE 4.9
Global trade in HCB and DDT.

4.3.7 Environmental Data

If this case study were a hypothetical classroom exercise or a laboratory experiment, we might have two sets of environmental data, "before" and "after," to bracket the regulatory restrictions on HCB neatly and demonstrate decreasing concentrations as a result of bans on its use. Instead, we have imperfect and incomplete evidence that still allows us to draw some general conclusions. These data are both time-limited (i.e., we do not have representative data collected at the peak of HCB use) and geographically limited. With respect to monitoring POPs in air and human blood and milk, for example, the United Nations has noted that data are available for central Europe, North America, and the Arctic but weak or unavailable in Africa, Asia-Pacific, Eastern Europe, and the Antarctic [119].

In 2005, a team of scientists [92] applied knowledge of the fate and transport of HCB to a detailed review of environmental data reported in the literature. These data included the concentrations of HCB in air, soil (background and treated agricultural soil), vegetation, seawater, plankton, and sediment samples taken from around the world. The scientists concluded that [92, p. 22]:

> The consistent downward trend [of concentrations] in the environment over the past 20 years suggests historic deposits of HCB are being depleted and HCB is being removed from the environment. Temporal trends of HCB in the environment vary, dependent on time period measured, media studied and study location, with peaks occurring in the 1960s, 1970s, and 1980s. The only places that do not have a consistent downward trend are in China, where HCB is still manufactured, in certain polar locations and in estuaries of historical source regions, which are presumably affected by catchment run-off and remobilisation of contaminated sediments. Marine mammals also often show no consistent upwards or downwards trends.

In short, this analysis of global data gave us qualified good news. Two of the caveats, regarding conditions in China and in certain polar regions, are worth a closer look.

A survey of HCB data for various media in China [89] found that HCB levels in air samples in the 2000s were higher than present-day concentrations in the European Union and the United States but similar to levels in those regions in the 1990s. This finding was attributed to the later production and prohibition of HCB in China. Few data are available to illustrate temporal trends in China [89]. The concentrations of HCB in air samples from Hong Kong were similar in the winter of 2000–2001 and 2005–2006. HCB concentrations in the sediments of Naisihu Lake in Shandong province increased significantly over the period 1960–2000. The concentrations of HCB in tissues of finless porpoises in the South China Sea did not change significantly in the last decade. Finally, HCB levels in human milk decreased in Beijing from 1988 to 2002 and again from 2002 to 2009–2010, and in Changchun city from 1987 to 1998, but increased by 60% in Changsha from 1988 to 1998. The observations that the concentrations decreased in some areas but increased in one region may relate to variations in point source emissions in those different places [89,120]. In short, it appears that the attenuation of HCB in China lags that in the west, where HCB bans eliminated many sources of HCB decades earlier. Conditions in the Arctic may reflect another kind of time lag.

As a result of concerns over the long-range transport of POPs to the Arctic, the Arctic Monitoring and Assessment Programme (AMAP) monitors the levels of POPs such as HCB in air, biota (fish, marine mammals, and seabirds), and humans. AMAP's 2014 report [121] indicates that:

- The concentrations of HCB at two Arctic air monitoring stations have increased in the last decade and declined very slowly at a third. Data from a fourth station in Greenland have not varied seasonally or correlated to ice cover, which may mean that HCB detected at that station originated from ongoing sources to the south (such as manufacturing by-products or incomplete combustion processes) rather than historical releases of HCB transported by the grasshopper effect.

- Trends in HCB concentrations in the tissue of freshwater fish, seabirds, marine mammals, marine fish, and mussels vary with species. Approximately half of the species monitored since before 2000 show no concentration trends, and decreasing trends were observed for many of the remaining species. However, when the analysis focused on data collected since 2000 it showed increasing trends for some species.
- HCB is commonly detected in the blood of people living in northern Canada, Finland, Greenland, Russia, and the United States (Alaska). Few time-series data are available. A study of Yupik mothers in Alaska that compared HCB levels in blood samples collected between 1999–2003 and 2004–2006 found that the concentrations were increasing with time; a similar study with Inupiat mothers found that the levels of HCB in blood decreased between 1997 and 1999–2003. These two groups of women lived in different areas and ate different diets, which could account for different exposures.

HCB has been detected in most of the environmental or biological samples collected for analysis around the world. As described above, bans of HCB that began in the 1980s appear to be gradually resulting in decreased environmental concentrations with a few exceptions. China, which banned the production and use of HCB some 20 years after the United States and the European Union, has not yet seen a substantial reduction in environmental concentrations, although the data are sparse. The long-range transport of HCB to remote Arctic regions apparently continues and because HCB can bioaccumulate, it is found in many animals and in humans. Those levels are not consistently decreasing despite the restrictions on the use of HCB.

4.3.8 Synthesis

The story related here about HCB is one of consequences, unintended and intended. The scientists who recommended seed treatment with HCB likely had the best of intentions. They recognized that wheat smut could easily destroy half a wheat crop. The new and very remarkable compound HCB could avert famine. But those scientists could not foresee that the compound's high vapor pressure and resistance to degradation would result in global transport of HCB, or anticipate the possible consequences of exposure. Alarms began to sound in the late 1940s about the possible dangers of exposure to pesticides, such as this editorial in the *Journal of the American Medical Association* in 1948 [122, pp. 1604–1605]:

> The development and use of new pesticides and herbicides have created public health hazards. Studies indicate that already many constituents

> of the American diet may be contaminated with materials of unknown toxicity used to destroy insect pests or weeds. …
>
> The brands of new insecticides, fungicides and herbicides (weed killers), now on the market or contemplated, number several thousand. The rapidity with which this industry has added new basic materials since the introduction of DDT is phenomenal. The Council on Pharmacy and Chemistry and the Council on Foods and Nutrition are deeply concerned over this situation because so little is known about either the acute or chronic pathologic effects on man. …

But there was no mechanism to explore such alarms or to take action on them. The writer went on to say "[a]t present governmental agencies find it difficult to protect the public from this potential hazard; existing agencies seem to doubt their authority and lack the funds to establish control" (pp. 1604–1605). The science of regulatory toxicology was in its infancy and environmental regulations minimal when HCB came to market. In the United States, for example, the Federal Insecticide, Fungicide and Rodenticide Act passed in 1947 did not require toxicity testing of pesticides, and the manufacture and use of nonpesticidal chemicals were not regulated until the 1970s.

Two incidents—HCB poisoning in Turkey in the 1950s and contamination of cows in the United States in 1972—raised awareness of the possible hazards from HCB. Each case was an unintended consequence of the manufacture and use of HCB that reflected both our ignorance and our lack of regulations. Current laws in most countries would require labeling of treated seed but hungry people may ignore those labels. Pesticide poisoning from eating treated seed happens even now [123]. Current environmental regulations in many countries also place tighter limits on the disposal of industrial waste than existed in 1972 and would prevent the open-air disposal of HCB residuals in an unregulated dump that led to the contamination of cattle in 1972.

This story also describes how intended consequences may sometimes fall short. Bans of HCB in the United States and the European Union promulgated with the intent to halt environmental pollution seem to have reduced the concentrations of HCB in air, water, and soil in many places. But the long-range global transport of HCB from incidental releases appears to continue and monitoring results from the Arctic suggest that contamination in remote areas is slow to diminish. The Stockholm Convention, nearly 10 years in the making from the initial proposal to final ratification, was intended to eliminate the production, use, import, and export of the "dirty dozen" chemicals. But the effectiveness of the treaty is not clear. Although it spurred China to cease production of HCB, data on the global trade of HCB/DDT suggest that the Stockholm Convention has had little effect on trade.

The next case study provides another perspective on unintended consequences.

4.4 Microbeads

Microplastic fragments less than 1 millimeter (mm) in size float in the oceans. They have washed up on shores around the world [124] and been found in sediments deep below the oceans' surface [125]. These tiny plastic particles originate from many sources. Lost fishing nets and discarded debris weather into fragments; washing clothing made from synthetic fabrics generates minute plastic fibers that pass through sewage treatment plants; virgin plastic pellets lost before they entered any manufacturing process attrit into smaller pieces [126,127]. Microplastics enter the environment from these sources as an indirect consequence of the use or disposal of a wide variety of plastic materials. One form of microplastic, however, is intended for a use that releases it directly to wastewater and often, consequently, to surface water. This case study examines the use and environmental fate of plastic microbeads.

4.4.1 Introduction

"This invention relates to a novel skin cleaner and more particularly to skin cleaners having plastic synthetic resin material in a fine state of comminution included therein so as to impart a scrubbing or mechanical detersive action thereto." So begins US Patent 3,645,904 [128]. Filed in 1967 and granted in 1972, this patent led to a host of skin care products that contained tiny plastic beads.

As we explore the use of these microbeads and the resulting consequences, we'll examine data pertaining to three kinds of products.

- Microbeads: Spheres made of polyethylene or another form of plastic, typically up to 300 microns in diameter.
- Microplastics: Plastic pieces less than 1 millimeter in size, commonly generated from fragmentation of larger pieces of plastic waste and sometimes including microbeads.
- Plastic spheres: Scientists have collected environmental and ecotoxicity data on plastic spheres larger in size than microbeads (e.g., up to 1300 microns in diameter).

4.4.2 Physicochemical Characteristics

Table 4.7 provides information on the composition of microbeads used in skin cleansers.

4.4.3 Manufacture and Use

Table 4.7 indicates the amounts at which microbeads may be incorporated into skin cleaners based on US patents. Analysis of two cleansers

TABLE 4.7
Composition of Microbeads

Composition	Specific Gravity	Particle Size (Microns)	Amount Used in Cleanser (wt% of Total)	Reference
Polyethylene (low density [LDPE], medium density, high density [HDPE])	0.910–0.965	74–77 desirable; up to 250 acceptable	8–10%	[128]
Polypropylene (PP)	0.900–0.915		Not Specified	[128]
Polystyrene G.P.	1.04–1.07		4–8%	[128]
Expanded polymers or copolymers, preferably based on vinylidene chloride, acrylonitrile, vinyl chloride, acronyl or styrene-based monomer, or a mixture of these monomers	0.015–0.200	80–300	Optimum 0.3–1.5% ; range from 0.1% to 10%	[129]

commercially available in Canada circa 1991 found that plastic microbeads comprised approximately 7 wt% of one product and 47 wt% of another [130]; six hand cleaners and facial scrubs purchased in New Zealand a few years later contained between 0.19 and 6.91 wt% plastic micrograins [131]. Further analyses of commercial products have shown that a median particle size, based on samples of four water-based facial cleansers containing polyethylene and purchased in New Zealand, ranged from 197 to 375 micrometers [132]; analysis of a facial scrub available in Australia found that the microbeads were 11 to 700 micrometers in diameter and irregularly shaped [133].

Research for this book did not uncover any reliable estimates of the tonnage of microbeads used in cleansing products. Such products are or have been sold in many countries [130,132,134,135]. For example, as of October 2014, approximately 60 cosmetic and personal care products containing microbeads were reportedly on the market in the United States (down from over 100 previously), 53 in Germany, and 51 in Hong Kong [135,136].

4.4.4 Fate and Transport in the Environment

Plastic microbeads enter domestic sewage when people use cleansers that contain them and wash the residues down the drain. The paragraphs below describe the likely fate of microbeads in a typical wastewater treatment plant and then in the environment.

Wastewater treatment usually begins with screening to remove coarse solids such as branches, paper, and rags; microbeads, by virtue of their size, would pass through such initial treatment units. Even grit removal units, which are designed to remove smaller particles, would not likely trap microbeads. Aerated grit chambers can remove particles 2 mm or larger, and grit chambers operating by gravity settling are designed to remove particles with a specific gravity of 2.65 [137]. Most microbeads are too small and too light to be captured efficiently.

After these initial treatment steps, wastewater typically flows to a primary clarifier for additional solids removal, then to a bioreactor and a secondary clarifier (Figure 2.5). Neither a primary clarifier nor a secondary clarifier is likely to remove microbeads efficiently. Engineers design clarifiers to remove solids with a specific gravity of approximately 1.03 from domestic sewage by gravity settling [137]. Based on the particle densities indicated in Table 4.7, most microbeads would float and remain in the wastewater after clarification. Although it is possible that microbeads could be removed during primary clarification when flocculants are added, at least one authority on wastewater treatment believes they are not [138]. Biological treatment is unlikely to remove microbeads, as plastics biodegrade poorly. It is possible that some of the plastic mass might become enmeshed in or adsorbed to the biological floc and be removed with the sludge in the secondary clarifier.

Following biological treatment and clarification, treatment plants disinfect the water and many discharge it to surface water. Some treatment plants use advanced treatment methods in addition to the unit operations described here, and it is possible that some advanced treatment methods may remove microbeads [139].

Based on the characteristics of microbeads and this understanding of the operations of a typical sewage treatment plant, one would expect that much of the mass of microbeads used in cleansers would pass through wastewater treatment to be discharged into surface water. Once in surface water, plastic particles might float indefinitely. Carried by waves and ocean currents, buoyant microplastics have been found in surface waters far from land and have notoriously accumulated in the Great Pacific "Garbage Patch" at the center of the North Pacific gyre [140]. Scientists hypothesize that plastic fragments are also accumulating in less-studied gyres [141,142].

However, some data suggest that microbeads, unlike other microplastics, may not significantly accumulate in ocean waters. One recent study examined the data from over 3,000 surface water samples collected around the world. Eighty-eight percent of the samples from the open ocean contained plastic debris. Microbeads were common in samples from near shore environments but they were negligible in open ocean samples [142]. That finding suggests that the processes that may remove microbeads from surface waters deserve a closer look.

Surface water flow may transport microbeads until they degrade, are ingested by organisms, wash up on a shoreline, or perhaps, come to rest in sediments. Those fate and transport processes are described briefly below; the potential for organisms to ingest microbeads is discussed in the overview of (eco)toxicological concerns.

Three mechanisms may break down microbeads. First, they may fracture or mechanically abrade to form smaller particles, perhaps small enough to be missed in environmental sampling. Second, photocatalyzed oxidation can degrade plastics and may be the primary mechanism for the breakdown of plastics in the environment, although it can be retarded in seawater [127]. Photodegradation can reduce the tensile strength and elasticity of a plastic, making the pieces more brittle and prone to breakage [140]. Mechanical or photo-oxidation–enhanced breakdown of plastic particles can increase the relative surface area of the plastic and thereby enhance further degradation of the polymers. Finally, some plastics can slowly biodegrade. The results of laboratory and field studies illustrate just how slowly plastics may degrade. One study, for example, indicated a 10-year carbon conversion rate by biodegradation for LDPE of 0.2% and 5.7%, without and with prephotodegradation, respectively. In another study, scientists immersed pieces of HDPE, LDPE, and PP in the Bay of Bengal (India) for six months; they measured the weight loss of these samples as 1.5–2.5% for LDPE, 0.5–0.8% for HDPE, and 0.5–0.6% for PP [127].

Microplastics floating in surface water may wash up on shorelines. Scientists have found microplastics on sandy shores in six continents [124], including beaches in remote areas far from obvious sources [126]. However, no data quantifying the concentration of microbeads in beach litter were identified during the research for this book.

Some microbeads may ultimately be deposited in sediments. Although it is counterintuitive that microbeads which float through a sewage treatment plant could ultimately come to rest in sediments, scientists have proposed two possible transport mechanisms that could carry microplastics to a lake or ocean floor [125,142 144]. The first mechanism can be described as follows. Biopolymers exuded by certain organisms may aggregate small particles of living organisms and inert material in a process sometimes referred to as biofouling. The resulting aggregates can sink through the water column as "marine snow" at rates that may reach up to nearly 400 meters per day. These aggregates may degrade and rerelease plastic particles, however, this mechanism is thought to transport some plastic fragments into sediment. The second possible mechanism may occur after aquatic organisms ingest plastics: microplastics in feces may sink to the sediment. Microplastics have been found in sediments, perhaps as a result of one or both of these hypothesized mechanisms, however, no data regarding the deposition of microbeads into sediments were identified during the research for this book.

As microplastics move through the environment they can sorb and transport other contaminants. Plastic debris scooped from the ocean has contained polychlorinated biphenyls (PCBs) at concentrations up to 5 parts per million and dichlorodiphenyltrichloroethane (DDT) at approximately 7 PPM, for example [145–147]. Of particular concern are persistent, bioaccumulative, and toxic (PBT) compounds that would tend to partition into a plastic matrix due to their hydrophobicity; the logical inference is that floating microplastics could transport PBT compounds through the environment and that perhaps the PBT compounds would enter the food chain as organisms ingested the plastic fragments.

TABLE 4.8

Estimated Mass Transport to Arctic

	Annual Mass Transport		
Compound	Via Floating Plastics	Via Water	Via Atmospheric Transport (Tons)
PCBs	1 g–1.6 kg	5.7 tons	300
Polybrominated diphenyl ethers (PBDE)	25 g–5.9 kg	Not calculated	720

Thermodynamic calculations provide some perspective on this intuitive concern. Models of the sorption of PBT substances to polyethylene suggest two important conclusions [148]. First, compounds with log K_{ow} >5 may partition >1% to polyethylene. (This degree of sorption may represent the worst case with respect to sorption to plastics in general, as other work has shown that polyethylene can sorb organic contaminants more effectively than other plastics such as polypropylene and polyvinyl chloride [149].) Second, although thousands of tons of plastic debris float in the earth's oceans, the quantity of plastic is relatively small compared to the volume of air, water, and organic carbon; consequently, thermodynamic calculations indicate that most of the mass of hydrophobic pollutants will cycle between air and water, potentially sorbing to naturally occurring carbon, and only a relatively small proportion of the total mass will sorb to floating plastics.

A mass balance on the transport of two PBT substances to the Arctic by different vectors illustrates this second point. Table 4.8 shows the results of these calculations, which were based on measured concentrations in the environment [150]. These calculations indicate that winds and water currents likely transport many times the mass of PBT compounds that is carried by floating plastics.

Although this work did not focus specifically on microbeads, the studies described above came to two relevant conclusions: hydrophobic compounds can partition into plastic debris and the relative mass transported in that fashion may be limited. Section 4.4.5 further discusses (eco)toxicological concerns relative to these findings.

4.4.5 Overview of (Eco)Toxicological Concerns

Scientists caution about a different sort of hazard from exposure to microbeads than might result from exposure to chemical pollutants. Aquatic organisms can ingest microplastics, which can then accumulate in the organism, limiting nutrition, causing blockages, and potentially releasing chemicals from within the plastic. One team of researchers summarized these concerns as follows [132, p. 1228]:

> Microplastics of [this size] can be ingested by filter-feeding polychaetes, echinoderms, bryozoans, bivalves and barnacles, deposit feeding lugworms and sea cucumbers, and by detritovores such as amphipods. More disturbingly, [studies] have recently shown that microplastics accumulate in the gut of filter-feeding mussels, are translocated to the circulatory system within three days of ingestion, and persist for more than 48 days.
>
> The microplastics described here are polyethylene, which with a specific density <1 will float on the water surface … and be available to a wide variety of planktonic organisms feeding in the euphotic zone, as well as fish and seabirds that feed at the water surface. Microplastics are consumed by planktonic organisms and plastic microspheres (0.01–0.07 mm) are consumed in laboratory feeding trials of copepods and invertebrate larvae. … If microplastics are ingested by small planktonic organisms such as copepods, there is the potential for the plastic to pass to, and accumulate, at higher levels of the food chain.

It is possible that some organisms may be able to eliminate microplastics. The overarching concern, however, is that the plastics may accumulate within an organism [126,127,142,151].

Once an organism has ingested microplastics, the accumulated material may have a mechanical effect, possibly blocking the intestinal tract or causing a feeling of satiation that results in reduced food intake. A few studies also show that microplastics may be translocated within an organism. Although one study of mussels, for example, showed translocation of microspheres into the circulatory system, it did not show any apparent sublethal effect; another study showed an immune response in the digestive glands of mussels exposed to microplastics [126].

Many scientists have raised the alarm that ingested microplastics can release toxic chemicals within an organism, whether those chemicals were sorbed into the plastic from contaminated water or whether the substances were used to manufacture the plastic [126,127,152]. Recent studies have explored the potential for this to occur. Let's begin by looking at the results of a thermodynamic model.

A team of scientists adapted the fugacity model CEMC SFU Agro to assess the fate of chemicals within pisciveorous fish whose diet included a 10% microplastic component. The model indicated that bioaccumulation of chemicals would actually decline as a result of the sorption of hydrophobic contaminants into the plastic within the organism. In other words, the plastic particles would serve as a sink rather than a source of anthropogenic chemicals. For chemicals with a log K_{ow} between 6.5 and 7.5, the model predicted that the body burden concentration would decrease by more than 20% if the diet contained 10% microplastics [148].

However, the availability of sorbed substances under physiological conditions can be influenced by the chemical environment in the body in ways that may not be accurately represented by a simple model. Persistent organic

TABLE 4.9
Results of Mass Balance on POPs and Ingested Plastics

Type of Plastic	Relative Affinity for POPs	Closed System	Open System
Polyethylene	Higher	Simulated bioaccumulation decreases substantially due to presence of plastic	Model predicts decrease in steady-state bioaccumulation for log K_{ow} >5 when microplastics in system
Polystyrene	Lower	Marginal increase in bioaccumulation due to presence of plastic	Negligible effect from inclusion of microplastics in system

pollutants (POPs) such PFOA and DDT can desorb up to 30 times more rapidly from polyethylene in a gut surfactant, particularly under conditions simulating warm-blooded organisms, than in seawater [153]. Phenanthrene desorbs from polyethylene approximately 7 times faster in solution with a gut surfactant than in seawater [146]. In fact, the release of POPs from the plastic surface in the gut may be so rapid that diffusion through the polymer may be the rate-limiting step in mass transfer [154].

A second-generation model comprised a mass balance of uptake and loss processes that more accurately reflected physiological processes through calibration to experimental data [154]. This model examined the mass transfer of PCBs from plastics to lugworms (*Arenicola marina*) upon ingestion and by dermal contact in two environments: closed laboratory bioassay systems and open marine systems. The results depended in part on the type of plastic, as polyethylene has a greater affinity for POPs than polystyrene. Table 4.9 summarizes those results.

Upon reporting these results, however, the authors cautioned that "error propagation due to parameter uncertainty can be expected to be considerable." They went on to conclude that "the role of plastic in bioaccumulation of POPs is scientifically interesting but not very relevant from a risk assessment perspective. This, however, may be different for chemicals for which plastic is the main source of bioaccumulation such as additives leaching from microplastic" (p. 7818).

That concern about additives leaching from microplastics was addressed in a study examining the fate of plasticizers in ingested plastics. Using a biodynamic model, researchers explored the potential of leaching of nonylphenol (NP) and bisphenol A (BPA) from plastics in the intestinal tracts of the lugworm (*Arenicola marina*) and North Sea cod (*Gadus morhua*). They concluded that this was "not likely to constitute a relevant exposure pathway" [155].

Although models such as these can predict the fate of chemicals within an organism, bioassays and animal studies indicate what actually happens upon exposure. The results of testing on lugworms, amphipods, and birds are described below.

Several bioassays have tested the effects of placing lugworms in sediment containing contaminant-sorbed plastics. One such study examined the effects of exposure to sediment containing microscopic polyvinyl chloride (PVC) particles (5%) loaded with phenanthrene, nonylphenol, tetrabromodiphenyl ether, and triclosan [156]. Lugworms did accumulate nonylphenol and phenanthrene from PVC in their tissues, however, worms exposed to sand containing *lower* concentrations of these substances (than were sorbed to the PVC) accumulated a substantially *higher* mass of the substances. Pollutants sorbed to plastics were less bioavailable than pollutants sorbed to sand. In a similar series of experiments, lugworms also accumulated PBDE and triclosan from PVC. Eating microplastics reportedly reduced survival (Triclosan), feeding (Triclocsan and PBDE), and immunity (nonylphenol). Another study (which was used to calibrate the model described above) examined the effects of exposing lugworms to polystyrene microplastics containing PCBs [157]. Lugworms were exposed to sediments pre-equilibrated with varying levels of polystyrene (0 to 7.4%) and contaminated with PCBs at levels that were not expected to cause immediate toxic effects. The results showed that the lugworms' feeding activity decreased with the highest amount of plastic in the sediment; lugworms in the high-plastic substrates also lost more weight. Lugworms ingested plastic spheres between 400 and 1,300 microns in size, but the particles did not accumulate in the organisms. Increased PCB accumulation with exposure to plastics was only observed at the lowest plastics loading of 0.074%, and that at a level which the authors concluded was probably not relevant for the risk assessment of POPs.

Another team of researchers focused specifically on the effects of ingestion of microbeads. They isolated the spheres from a commercially available facial cleanser and suspended them in solution with polybrominated diphenyl ether (PBDEs). Amphipods (*Allorchestes compressa*) exposed to a suspension of microbeads ingested up to 45 particles per animal and evacuated them within 36 hours. The authors of the report concluded that the presence of microbeads apparently inhibited the uptake of PBDEs into amphipods, relative to uptake from solutions of the same concentration of PBDEs that did not include microbeads, potentially by making PBDEs less bioavailable. However, the amphipods took up a greater proportion of the more highly brominated congeners when microbeads were present. (The authors offered two possible explanations for that result: either the microbeads concentrated the more highly brominated congeners for ingestion, or else the more highly brominated congeners sorbed onto the glassware in the controls that lacked microbeads and thus the compounds were not available for ingestion in the controls [133].)

Scientists have also looked at the consequences of birds ingesting contaminated plastic. The results of these tests can be difficult to interpret because the birds' natural prey can also contain hydrophobic contaminants. Three studies have examined the bioaccumulation of chlorinated organic pollutants in marine birds known to have ingested plastic. The earliest report dates back to 1988. In that study, multivariate analyses were used to assess the determinants of PCBs, DDE, DDT, and dieldrin in the fat and eggs of female great shearwaters (*Puffinus gravis*). The levels of PCBs in the birds' fat and eggs correlated with the mass of ingested plastic; the levels of the other contaminants did not [158]. A more recent study explored whether a small number of streaked shearwater chicks (*Calonectris leucomelas*) ingesting contaminated plastic could accumulate PCBs. The results showed that "transfer of PCBs, especially lower chlorinated congeners, occurs from ingested plastics to the biological tissue of the organisms that intake the plastics" [149, p. 2042]. In a third study, 12 short-tailed shearwaters (*Puffinus tenuirostris*) accidentally caught during experimental fishing in the North Pacific Ocean contained a mean mass of plastics in their stomachs of 0.23 grams per bird. The mass of ingested plastic correlated positively with concentrations of lower-chlorinated PCB congeners in abdominal adipose tissue [159]. Shearwaters in the wild may also be exposed to PBDEs sorbed to plastic. By analyzing the PBDE concentrations in the birds' abdominal adipose tissue, their typical prey, and the plastics, the research team concluded that the ingestion of plastics might be contributing to the body burden of the more highly brominated congeners in the 12 short-tailed shearwaters described above. [160].

In summary, research shows that ingestion of microplastics can have physical effects on an organism. Microplastics might also sorb hydrophobic contaminants and scientists have raised the concern that the contaminants released from ingested plastics could bioaccumulate and perhaps have toxic effects. Studies of this point to date have yielded mixed results. Some models predict that in fact ingested plastics could sorb POPs already in the organism from other sources, and by serving as a sink rather than a source of contaminants, reduce bioaccumulation. Bioassays and field studies to date do not unequivocally support the broad conclusion that ingestion of plastics laden with hydrophobic contaminants significantly contributes to bioaccumulation. With these results to provide context, let's look at how the use of plastic microbeads is regulated.

4.4.6 Regulatory Status

Polyethylene (CAS 9002-88-4) is listed on the TSCA Inventory [38] as an existing chemical with no SNUR (although polymers are generally exempt from TSCA in the United States under regulations at 40 CFR 723.250). Thus TSCA places no restrictions on the use of polyethylene.

Further complicating a regulatory analysis is the fact that it is the form of the plastic, in small beads, that gives rise to ecotoxicological concerns. Thus it is worth considering whether polymers in this form might be regulated. A microbead may fit the definition of an article under TSCA (40 CFR 720.3), that is, "a manufactured item (1) which is formed to a specific shape or design during manufacture, (2) which has end-use function(s) depending in whole or in part upon its shape or design during end use, and (3) which has either no change of chemical composition during its end use or only those changes of composition that have no commercial purpose separate from that of an article, and that results from a chemical reaction that occurs upon end use of other chemical substances, mixtures, or articles." In general, TSCA exempts articles that contain chemical substances which are not intended to be released and have no separate commercial purpose. Careful readers perhaps noticed the equivocal phrasing that a microbead *may* fit the definition of an article. One might consider a microbead to be a particle. Under TSCA, particles are not articles regardless of shape or design, but the regulations do not define "particles."

Absent clear limitations under TSCA and recognizing environmental concerns over the consequences of microbead use, federal and state governments have proposed or promulgated regulations on their use. At the federal level, the Microbead-Free Waters Act of 2014 was introduced into the House of Representatives in June 2014. It would prohibit the sale or distribution of cosmetics containing synthetic plastic microbeads by 2018. The bill was referred to the House Committee on Energy and Commerce for consideration and no further action was taken in 2014 [161]. State legislatures also took notice of the issue. As of mid-2014, a bill to ban the use of microbeads had passed in Illinois, and bills were pending before the legislatures in California, New York, Michigan, and Minnesota. Illinois will phase in the ban between 2017 and 2019 [162].

In the European Union, polymers are generally exempt from registration requirements under REACH [163]. One could argue that microbeads used in cleansers are an "article" based on the definition of the term under REACH, that is, "an object which during production is given a special shape, surface or design which determines its function to a greater degree than does its chemical composition." REACH regulates articles only to the extent that they may release a chemical substance upon use (and the total amount of the substance manufactured or imported is >1 tonne per year) or else contain a Candidate List substance above a threshold level [164]. Neither provision would obviously pertain to microbeads in cleansers.

Certain Member States have expressed concerns over the use of microbeads in cleansers. The Dutch Parliament adopted a motion that requested that the Dutch government talk with industry stakeholders about avoiding the use of microplastics in cosmetics, and the government reportedly supports a ban on the use of microplastics in cosmetics based on the precautionary principle [165].

In China, polymers are generally considered to be of low concern and most existing polymers are not regulated [166]. Polyethylene (CAS 9002-88-4) is on the IECSC [41] as an existing chemical.

While many governments were assessing the possible risks from exposure to microbeads and the need to take action, "retail regulation" began to take effect. Nongovernmental organizations called attention to the environmental concerns regarding the use of microbeads; funded in part by the United Nations Environmental Programme, one coalition even created a smartphone application called "Beat the Microbead" that allows the user to scan a product's barcode to determine whether the product is in a database of products that contain microbeads [167].

In response to such campaigns, several multinational manufacturers of products containing plastic microbeads committed to phasing out their use. The timing of these commitments reflected the practical business reality that it takes time to reformulate products. Some companies committed to phasing out the use of microbeads within a year, others within a three-year period [168,169].

4.4.7 Environmental Data

Scientists are just beginning to collect data on the presence of microbeads in the environment. Much of that work has been done in and around the Great Lakes in North America. The work was ongoing while this book was being written; consequently, some of the information below originated in the "grey literature" rather than conventional peer-reviewed sources.

For readers outside North America, a bit of background information on the Great Lakes may be in order. Lakes Superior, Michigan, Huron, Erie, and Ontario lie on the border between the United States and Canada. Together these interconnected lakes hold roughly 18% of the world's fresh water. Over 33 million people live within the Great Lakes basin. Pollutants from human activities tend to be retained within the Great Lakes system, as less than 1% of the total volume of water flows out of the system every year [170].

The data trail on microbeads begins at the potential point of discharge to surface water. A research team sampled the effluent from three wastewater treatment plants in northeastern New York to determine whether the discharge contained microplastics such as microbeads. (Although these plants do not discharge directly to the Great Lakes, the data are germane to sources that do so.) The team found that the plants discharged on the order of 10^5 to 10^6 microplastic particles per day, apparently comprising both microbeads and fibers from clothing [171].

In the summer of 2012 an expedition set out to learn whether such sources could have discharged microplastics into the Great Lakes [172]. Sailing in Lakes Superior, Huron, and Erie, scientists collected 21 discrete samples of floating debris by towing a net with 0.333 mm openings behind their research vessel. (It is worth noting that, based on the size information presented in Sections 4.4.2 and 4.4.3, some microbeads could pass through this mesh.) Back in the laboratory, the samples were sieved and the fractions less than 0.999 mm in size examined using scanning electron microscopy (SEM)/ Energy Dispersive X-Ray Spectroscopy (EDS) to evaluate surface characteristics and chemical composition.

Twenty of the samples contained plastic at levels of up to 466,305 pieces per square kilometer of surface water. Samples from Lake Erie, particularly from the New York area, were more concentrated than samples from Lakes Huron or Superior. Particles of a size between 0.355 and 0.999 mm comprised 81% of the total particle count. The research team concluded that [172, p. 5]:

> SEM/EDS analyses of all particles <1 mm differentiated microplastics, including multi-colored plastic spheres, from other materials such as coal ash. Based on dense urban populations adjacent to the lakes that employ combined sewage overflow, and the convergence of lake currents near our sample sites, we believe the microplastic spheres may be microbeads used in consumer product applications, such as those used in facial cleansers.

A second expedition in 2013 collected 91 samples from Lakes Michigan, Erie, and Ontario. The samples reportedly contained "significant amounts" of microbeads [136].

Microbeads have received attention due to the concern that fish may ingest them. In a 2013 diet assessment study, scientists examined the stomachs of nearly 1,000 fish from Lake Erie using a microscope to search for microplastics. Based on a very preliminary report of that work, they did not detect microbeads in the samples [173]. The work continues as of this writing in late 2014. Another study, of fish in Lake Superior, reportedly did not find plastic in the fish samples tested [174].

4.4.8 Synthesis

This case study described the use of microbeads in cleansing products and concerns over the consequences of exposure to microbeads. By design, the use of these products releases plastic particles to wastewater and then to surface waters. Scientists have raised concerns that these plastic microbeads do not readily degrade and may be ingested by aquatic organisms. No data identified as of this writing confirm this hypothesis that microbeads are ingested. Although study results are mixed on the significance of these effects, microbeads may sorb hydrophobic compounds and transport them through the environment or leach those compounds upon ingestion.

Chemical control regulations generally do not impose restrictions on the polymers used to manufacture microbeads and in any event it is the form of the product—in tiny particles that could be ingested—rather than the chemical composition which gives rise to concerns. With the clarity of hindsight, it is curious that ecological concerns over the use of microbeads only arose some 40 years after their use began. Consider the timeline shown in Figure 4.10.

Evidence that microplastics could resemble plankton, at the base of the food chain, surfaced early. Scientists first collected plastic fragments in plankton samples in the 1960s [175]. By the early 1970s scientists had observed 0.5 mm plastic spherules floating in the ocean, "widespread in the coastal waters of southern New England" in the United States and in the Sargasso Sea. The plastic spherules found in New England waters had sorbed PCBs and were ingested by fish [176,177]. These polystyrene pellets were unrelated to the polyethylene microbeads later used in cleansers, however, they provided a remarkable model for later concerns over the behavior of microbeads.

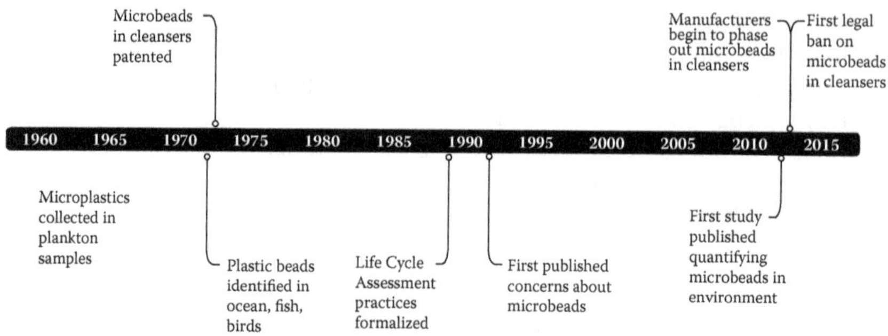

FIGURE 4.10
Timeline: manufacture and regulation of microbeads.

Other early studies pointed to a possible effect on the food chain. Two scientists wrote in 1974 [178, p. 44]:

> Small polystyrene particles, evidently of industrial origin, now appear as a contaminant of the sea in several parts of the world. They have been discovered in pellets of indigestible food regurgitated by gulls and terns, so are clearly entering the food chain at some point. So far as is known at present, they are harmless but it would be as well to exercise caution in releasing plastic to the environment.

These early concerns in the scientific community paralleled the reasons that governmental and nongovernmental organizations later used to call for a ban on products containing microbeads. However the first published warning about the possible consequences of using microbeads in cleansers did not appear until 1991 [130], and governmental actions first occurred roughly 20 years after that.

In hindsight, it may seem obvious that using microbeads in cleansers might have environmental effects that should have been considered in manufacturing and purchasing those products. But the development of cleansers containing microbeads predated the formal practice of life cycle assessment by some 20 years, and absent regulatory controls or the public dissemination of information published in scientific journals, no one seemed to ask for many years whether using microbeads in cleansers made environmental sense. Meanwhile, driven by consumer demand, over 100 products containing plastic microbeads reached the market. One key piece of data needed to assess the possible environmental consequences of using these products, the tonnage of plastic microbeads used over time, is not readily available. Chemical reporting rules such as TSCA CDR do not apply and there's a competitive disadvantage to a manufacturer voluntarily publishing that information.

Based on "retail regulation" and emerging regulatory actions by various governmental bodies, many manufacturers have committed to phasing microbeads out of their products. One might conclude that this demonstrates how the powerful combination of scientific evidence and public activism can limit global environmental pollution but for a curious postscript.

Scientists searching for evidence of microbeads in the environment have also found evidence of plastic fibers from our clothing. Consider this [124, p. 9175]:

> Experiments sampling wastewater from domestic washing machines demonstrated that a single garment can produce >1900 fibres per wash. This suggests that a large proportion of microplastic fibres found in the marine environment may be derived from sewage as a consequence of washing of clothes.

This finding is nothing new. In 1998, scientists wrote [179, p. 1]:

> Synthetic fibers are abundant in sludge, sludge products, and sewage treatment plant effluents. The fibers evidently are introduced from clothes-washing machines and survive the sewage treatment process.

As noted above, a more recent study of the effluent from wastewater treatment plants in New York found that the plants discharged up to a million microplastic particles per day, apparently comprising both microbeads and fibers from clothing [171]. Examination of the data from over 3,000 surface water samples collected around the world showed that "textile fibers are common in near shore environments" [142]. Analysis of microplastics in sewage sludge disposed of at sea found that polyester (67%) and acrylic (17%) fibers predominated, which suggested that the microplastics originated primarily from washing clothing and not from microbeads [124].

Such data raise the obvious question of why we view microbeads and textile fibers differently as environmental pollutants. An unscientific Google search turns up hundreds of references to banning microbeads, many quite passionate, but only a handful on synthetic fabrics. Why the difference?

References

[1] Green, R.S. 1963. Data gathering and monitoring equipment in water supply and water pollution control programs. *Amer. J.Public Health . Nation's Health.* 53(12): 1963–1971.

[2] Middleton, F.M., Lichtenberg, J.J., and Taft, R.A. 1960. Industrial Wastes—Measurements of organic contaminants in the nation's rivers. *Indust. Eng. Chem.* 52(6): 99A–102A.

[3] US Department of Health, Education, and Welfare – Public Health Service. 1959. *Report on the Recovery of Orthonitrochlorobenzene from the Mississippi River.* June 22.

[4] Howard, P.H., Santodonato, J., Saxena, J., Malling, J., and Greninger, D. 1976. *Investigation of Selected Potential Environmental Contaminants: Nitroaromatics*. EPA-560/2-76-010. US EPA Office of Toxic Substances.

[5] Council on Environmental Quality (CEQ). 1971. *Toxic Substances*, reprinted in Staff House Committee on Interstate and Foreign Commerce, 94th Congress, 2nd Session, Legislative History of the Toxic Substances Control Act (TSCA Legislative History) at 760 (Comm. Print 1976), p. 776.

[6] Yalkowsky, S.H. and Dannenfelser, R.M. 2008. *Aquasol Database of Aqueous Solubility. Version 5. College of Pharmacy, University of Arizona.* As cited in: US National Institutes of Health (NIH), National Library of Medicine (NLB). 1-Chloro-2-Nitrobenzene. CASRN: 88-73-3. Hazardous Substances Databank (HSDB) Number: 1322. Last Revision Date: 20080422. Available at: http://toxnet.nlm.nih.gov/cgi-bin/sis/search/a?dbs+hsdb:@term+@DOCNO+1322 (accessed October 18, 2013).

[7] Altschuh, J. et al. 1999. *Chemosphere* 39: 1871–1897. As cited in: US NIH, NLM. 1-Chloro-2-Nitrobenzene. CASRN: 88-73-3. HSDB No.: 1322. Last Revision Date: 20080422. Available at: http://toxnet.nlm.nih.gov/cgi-bin/sis/search/a?dbs+hsdb:@term+@DOCNO+1322 (accessed October 18, 2013).

[8] MITI. 1992. *Biodegradation and Bioaccumulation: Data of Existing Chemicals Based on the CSCL Japan*, Compiled under the Supervision of Chemical Products Safety Division, Basic Industries Bureau MITI, Ed. by CITI. Published by Japan Chemical Industry Ecology-Toxicology & Information Center. As cited in: United Nations Environmental Programme (UNEP) Chemicals, 2001. SIDS Initial Assessment Report for SIAM 13. 1-Chloro-2-nitrobenzene. Available at: www.chem.unep.ch (accessed May 2014).

[9] US NIH, NLM. *1-Chloro-2-Nitrobenzene. CASRN: 88-73-3. HSDB No.: 1322*. Last Revision Date: 20080422. Available at: http://toxnet.nlm.nih.gov/cgi-bin/sis/search/a?dbs+hsdb:@term+@DOCNO+1322 (accessed October 18, 2013).

[10] Hansch, C., Leo, A., and Hoekman, D. 1995. *Exploring QSAR - Hydrophobic, Electronic, and Steric Constants*. Washington, DC: American Chemical Society, p. 17. As cited in: US NIH, NLM. 1-Chloro-2-Nitrobenzene. CASRN: 88-73-3. HSDB No.: 1322. Last Revision Date: 20080422. Available at: http://toxnet.nlm.nih.gov/cgi-bin/sis/search/a?dbs+hsdb:@term+@DOCNO+1322 (accessed October 18, 2013).

[11] National Institute of Technology and Evaluation. 2007. *Japan Chemical Industry Ecology - Toxicology and Information Center*. Available at: http://www.safe.nite.go.jp/english/kizon/KIZON_start_hazkizon.html as of July 2, 2007. As cited in: US NIH, NLM. 1-Chloro-2-Nitrobenzene. CASRN: 88-73-3. HSDB No.: 1322. Last Revision Date: 20080422. Available at: http://toxnet.nlm.nih.gov/cgi-bin/sis/search/a?dbs+hsdb:@term+@DOCNO+1322 (accessed October 18, 2013).

[12] Yalkowsky, S.H. and He, Y. 2003. 1,4-dichlorobenzene and 1,3-dichlorobenzene. In: *Handbook of Aqueous Solubility Data*. Boca Raton, FL: CRC Press, pp. 207–208. As cited in: Agency for Toxic Substances and Disease Registry (ATSDR), 2006. Toxicological Profile for Dichlorobenzenes. August. Available at: http://www.atsdr.cdc.gov/toxprofiles/tp.asp?id=704&tid=126 (accessed January 15, 2014).

[13] Shiu, W.-Y. and Mackay, D. 1997. Henry's law constants on selected aromatic hydrocarbons, alcohols, and ketones. *J. Chem. Eng. Data*. 42:27–30. As cited in: ATSDR, 2006. Toxicological Profile for Dichlorobenzenes. August. Available at: http://www.atsdr.cdc.gov/toxprofiles/tp.asp?id=704&tid=126 (accessed January 15, 2014).

[14] US EPA. 2012. *EPI Suite™ v4.11. EPISuite User Guide*. Available at: http://www.epa.gov/oppt/exposure/pubs/episuitedl.htm (accessed January 4, 2014).
[15] Hansch, C., Leo, A., and Hoekman, D. 1995. o-Dichlorobenzenes and p-dichlorobenzenes. In *Exploring QSAR. Hydrophobic, Electronic, and Steric Constants*. Washington, DC: American Chemical Society. As cited in: ATSDR, 2006. Toxicological Profile for Dichlorobenzenes. August. Available at: http://www.atsdr.cdc.gov/toxprofiles/tp.asp?id=704&tid=126 (accessed January 15, 2014).
[16] Wilson, J.T. et al. 1981. *J. Environ. Qual.* 10:501–506. As cited in: As cited in: US NIH, NLM. 1,4-Dichlorobenzene. CASRN: 106-46-7. HSDB No.: 523. Last Revision Date: 20081007. Available at: http://toxnet.nlm.nih.gov/cgi-bin/sis/search/a?dbs+hsdb:@term+@DOCNO+523 (accessed December 8, 2013).
[17] Oliver, B.G. and Niimi, A.J. 1983. Bioconcentration of chlorobenzenes from water by rainbow trout: Correlations with partition coefficients and environmental residues. *Environ. Sci. Technol.* 17: 287–291. As cited in: ATSDR, 2006. Toxicological Profile for Dichlorobenzenes. August. Available at: http://www.atsdr.cdc.gov/toxprofiles/tp.asp?id=704&tid=126 (accessed January 15, 2014).
[18] Yalkowsky, S.H. 1992. *Aquasol Database of Aqueous Solubility*. 5th edition. Tucson: University of Arizona, College of Pharmacy. As cited at: ATSDR, 2013. Toxicological Profile for Hexachlorobenzene. Draft for Public Comment - June. Available at: http://www.atsdr.cdc.gov/toxprofiles/tp.asp?id=627&tid=115 (accessed January 14, 2014).
[19] Warner, H.P., Cohen, J.M., and Ireland, J.C. 1987. *Determination of Henry's law constants of selected priority pollutants*. PB-87-212684/XAB; EPA-600/D-87/229. Environmental Protection Agency, Cincinnati, OH. Water Engineering Research Laboratory. (accessed January 14, 2014).
[20] United Nations Industrial Development Organisation. Undated. *Hexachlorobenzene - Environmental Impact*. Available at: http://www.unido.org/en/what-we-do/environment/capacity-building-for-the-implementation-of-multilateral-environmental-agreements/the-stockholm-convention/facts-and-figures/what-are-persistent-organic-pollutants-pops/hexachlorobenzene/environmental-impact.html (accessed October 4, 2014).
[21] National Research Council. 2009. *Global Sources of Local Pollution: An Assessment of Long-Range Transport of Key Air Pollutants to and from the United States*. Washington, DC: National Academies Press.
[22] Hansch, C., Leo, A., and Hoekman, D. 1995. *Exploring QSAR: Hydrophobic, Electronic, and Steric Constants*. Washington, DC: American Chemical Society, 15. As cited in: ATSDR, 2013. Toxicological Profile for Hexachlorobenzene. Draft for Public Comment - June. Available at: http://www.atsdr.cdc.gov/toxprofiles/tp.asp?id=627&tid=115 (accessed January 14, 2014).
[23] Chiou, C.T., Porter, P.E., and Schmedding, D.W. 1983. Partition equilibria of nonionic organic compounds between soil organic matter and water. *Environ. Sci. Technol.* 17: 227–231. As cited in: ATSDR, 2013. Toxicological Profile for Hexachlorobenzene. Draft for Public Comment - June. Available at: http://www.atsdr.cdc.gov/toxprofiles/tp.asp?id=627&tid=115 (accessed January 14, 2014).
[24] Veith, G.D., DeFoe, D.L., Bergstedt, B.V., et al. 1979. Measuring and estimating the bioconcentration factor of chemicals in fish. *J. Fish Res. Board Can.* 36: 1040–1048. As cited in: ATSDR, 2013. Toxicological Profile for Hexachlorobenzene. Draft for Public Comment - June. Available at: http://www.atsdr.cdc.gov/toxprofiles/tp.asp?id=627&tid=115 (accessed January 14, 2014).

[25] Dugal, M. 2005. *Nitrobenzene and Nitrotoluenes. Kirk-Othmer Encyclopedia of Chemical Technology*. Hoboken: NJ: John Wiley & Sons. Published online: 14 Oct 2005.

[26] Bucher, J.R. 1993. *NTP Technical Report on Toxicity Studies of 2-Chloronitrobenzene and 4-Chloronitrobenzene* (CAS Nos. 88-73-3 and 100-00-5). NIH Publication 93-3382. US Department of Health and Human Services, National Institutes of Health. 1993. Available at: http://ntp.niehs.nih.gov/ntp/htdocs/st_rpts/tox33.pdf. P. 19.

[27] United Nations Environmental Programme (UNEP) Chemicals. 2001. *Screening Information Data Set (SIDS) Initial Assessment Report for SIAM 13*. 1-Chloro-2-nitrobenzene. Available at: www.chem.unep.ch (accessed May 2014).

[28] US International Trade Commission. 1995. *Synthetic Organic Chemicals – United States Production and Sales*, 1994. USITC Publication 2933. Seventy-Eighth International Edition.

[29] US EPA. 2014. *Chemical Data Access Tool (CDAT)*. Available at: http://java.epa.gov/oppt_chemical_search/(accessed July 14, 2013).

[30] IARC. 1996. *Monographs on the Evaluation of Carcinogenic Risks to Humans*. Volume 65: Printing Processes and Printing Inks, Carbon Black and Some Nitro Compounds. 2-Chloronitrobenzene, 3-Chloronitrobenzene, and 4-Chloronitrobenzene. pp. 263-296. Available at: http://monographs.iarc.fr/ENG/Monographs/vol65/index.php (accessed June 3, 2014).

[31] Ding, Y. 1999. Demand of o-nitrochlorobenzene to have a continuous increase. *China Chemical Reporter*. May 26. China National Chemical Information Center. Available at: http://www.ccr.com.cn (accessed May 5, 2014).

[32] Anhui Bayi Chemical Industry Co.,Ltd. 2014. *About Us*. Available at: http://www.bayichem.com/gsjj-en.html (accessed June 3. 2014).

[33] Solutia, Inc. 2004. *High Production Volume (HPV) Chemical Challenge Program – Test Plan for the Chloronitrobenzene Category*. Available at: http://www.epa.gov/hpv/pubs/summaries/chlrnbnz/c14392tc.htm (accessed May 30, 2014).

[34] Stockholm Convention. *Annex D: Information Requirements and Screening Criteria*. Available at: http://chm.pops.int/TheConvention/Overview/TextoftheConvention/tabid/2232/Default.aspx (accessed February 7, 2014).

[35] OECD. 2009. *OECD Pov and LRTP Screening Tool, Version 2.2*. Available at: http://www.oecd.org/chemicalsafety/risk-assessment/oecdpovandlrtpscreeningtool.htm (accessed September 12, 2014).

[36] Scheringer, M., MacLeod, M., and Wegmann, F. 2009. *Manual - The OECD POV and LRTP Screening Tool, Version 2.2*.

[37] Jones, C.R., Liu, Y.Y., Sepai, O., Yan, H., and Sabbioni, G. 2006. Internal exposure, health effects, and cancer risk of humans exposed to chloronitrobenzene. *Environ. Sci. Technol*. 40(1): 387–394.

[38] US EPA. 2014. *TSCA Chemical Substances Inventory*. File last created 02/24/2014. Available at: http://www.epa.gov/oppt/existingchemicals/pubs/tscainventory/howto.html (accessed May 5, 2014).

[39] Solutia, Inc. 2004. *High Production Volume (HPV) Chemical Challenge Program – Test Plan for the Chloronitrobenzene Category*. Available at: http://www.epa.gov/hpv/pubs/summaries/chlrnbnz/c14392tc.htm (accessed May 30, 2014).

[40] European Chemicals Agency (ECHA). 2014. *Registered Substances*. Available at: http://echa.europa.eu/web/guest/information-on-chemicals/registered-substances (accessed May 2, 2014).

[41] CIRS. 2014. *Chinese Chemical Inventory Search System (CCISS)*. Available at: http://cciss.cirs-group.com/(accessed May 31, 2014).

[42] US Department of Health, Education, and Welfare. 1962. *Public Health Service Water Pollution Surveillance System - Annual Compilation of Data, October 1, 1962 - September 30, 1963*. Volume 8: Southwest-Lower Mississippi River Basin.

[43] Illinois Department of Public Health. 2009. *Public Health Assessment, Sauget Area 2 Landfill Sites P, Q and R, Sauget, St. Clair County, Illinois*. EPA Facility ID # ILD000672329. Prepared on behalf of the Agency for Toxic Substances and Disease Registry. Available at: http://www.atsdr.cdc.gov/HAC/pha/pha.asp?docid=605&pg=1 (accessed May 2014).

[44] United States Geological Survey. 2012. *Sediment-Contaminant Database for the Upper Mississippi River System and Selected Tributaries (Version 2)*. Available at: http://www.umesc.usgs.gov/data_library/sediment_contaminants/sediment_contaminant_page.html (accessed May 31, 2014).

[45] Yurawecz, M.P. and Puma, B.J. 1983. Identification of chlorinated nitrobenzene residues in Mississippi river fish. *J. Assoc. Off. Anal. Chem.* 66: 1345–1352.

[46] Commission of the European Communities.1993. *Updating of Data Concerning the Impact on the Aquatic Environment of Certain Dangerous Substances*, 2nd part. Part III Chloronitrobenzenes and 4-chloro-2-nitroaniline. Available at: http://monographs.iarc.fr/ENG/Monographs/vol65/mono65-7.pdf (accessed June 4, 2014).

[47] OECD. 2003. HPV Chemicals Programme, SIDS Dossier, *1-Chloro-2-nitrobenzene CAS: 88-73-3*. Approved at SIAM 13 (6-9 November 2001). Date of last Update: 26-NOV-2003. Available at: http://www.inchem.org/documents/sids/sids/CHLORONITROB.pdf (accessed July 12, 2014).

[48] He, M., Sun, Y., Li, X., and Yang, Z. 2006. Distribution patterns of nitrobenzenes and polychlorinated biphenyls in water, suspended particulate matter and sediment from mid- and down-stream of the Yellow River (China). *Chemosphere*. 65: 365–374.

[49] US EPA. 2014. United States Environmental Protection Agency Regions 3, 6, and 9, 2014. *Regional Screening Levels for Chemical Contaminants at Superfund Sites*. Available at: http://www.epa.gov/reg3hwmd/risk/human/rb-concentration_table/index.htm (accessed September 14, 2014).]

[50] van Rumker, R., Lawless, E.W., and Meiners, A.F. 1975. *Production, Distribution, Use and Environmental Impact Potential of Selected Pesticides*. EPA-540/1-74-001, pp. 8; 320–326; 354.

[51] Krishnamurti, R. 2004. *Chlorobenzenes*. 6: 211–225. *Kirk-Othmer Encyclopedia of Chemical Technology*. 5th edition. Hoboken, NJ: John Wiley & Sons.

[52] Mayo, D., Collins, J., and Riordan, B. 1980. *Economic Impact Analysis of Proposed Test Rule for Chloromethane and Chlorobenzenes - Support Document for Proposed Health Effects Test Rule Toxic Substances Control Act Section 4*. Prepared for US EPA Office of Pesticides and Toxic Substances. EPA 560/11-80-21.

[53] Ware, S.A. and West, W.L. 1977. *Investigation of Selected Potential Environmental Contaminants: Halogenated Benzenes*. EPA 560/2-77-004. US EPA Office of Toxic Substances. July. Pp 1, 24-26, 66-67.

[54] Slimak, K., Johnston, P., and Hodge, V. 1979. *Materials Balance for Chlorobenzenes – Level 1*. EPA-560/13-80-001. Prepared for US EPA Office of Toxic Substances. December 31.

[55] ATSDR. 2006. *Toxicological Profile for Dichlorobenzenes.* August. Available at: http://www.atsdr.cdc.gov/toxprofiles/tp.asp?id=704&tid=126 (accessed January 15, 2014).

[56] US EPA. 1986. *Certain Chlorinated Benzenes; Final Test Rule.* 51 FR 11728 – 11737. April 7.

[57] European Commission. 2004. *European Union Risk Assessment Report - 1,4-Dichlorobenzene.* CAS No: 106-46-7. EINECS No: 203-400-5. Risk Assessment. Available at: http://echa.europa.eu/documents/10162/fb7bf6b4-7831-4c3b-87b3-5acd493ce597 (accessed December 8, 2013).

[58] US EPA. Undated. *Technical Factsheet on: p-DICHLOROBENZENE.* Available at: http://water.epa.gov/drink/contaminants/basicinformation/historical/upload/Archived-Technical-Fact-Sheet-on-p-Dichlorobenzene.pdf (accessed August 20, 2014).

[59] Risk & Policy Analysts Limited. 2010. *Socio-Economic Evaluation Arising from a Proposal for Risk Reduction Measures Related to Restrictions on 1,4 Dichlorobenzene.* Final Report Prepared for European Commission Directorate-General Enterprise and Industry. Available at: http://ec.europa.eu/enterprise/sectors/chemicals/files/docs_studies/final-report-1-4-dichlorobenzene_en.pdf (accessed August 19, 2014).]

[60] Anonymous. 1996. A Glimpse at the p-Dichlorobenzene Market. *China Chemical Reporter.* October 6. Available at: http://www.highbeam.com/doc/1G1-46776079.html (accessed August 19, 2014).

[61] National Library of Medicine. 2008. *1,4-Dichlorobenzene - National Library of Medicine Hazardous Substances Database (HSDB).* Last Revision Date: 20081007. Available at: http://toxnet.nlm.nih.gov/cgi-bin/sis/search/a?dbs+hsdb:@term+@DOCNO+523 (accessed August 19, 2014).

[62] US EPA. 2014. *Chemical Data Access Tool (CDAT) – 2012 CDR Data.* Last updated June 2014. Available at: http://java.epa.gov/oppt_chemical_search/(accessed August 19, 2014).

[63] United Nations. 2014. *United Nations Commodity Trade Statistics Database.* Last update in UNdata: data as of 1 Apr 2014. Available at: http://data.un.org/ (accessed August 2014).

[64] US Department of Health and Human Services. 2014. *Household Products Database - Chemical Name: p-Dichlorobenzene.* Available at: http://householdproducts.nlm.nih.gov/cgi-bin/household/brands?tbl=chem&id=1077&query=dichloro benzene&searchas=TblChemicals (Searched September 6, 2014).

[65] ECHA. 2013. *Committee for Risk Assessment (RAC)/Committee for Socio-economic Analysis (SEAC) - Background Document to the Opinion on the Annex XV Dossier Proposing Restrictions on 1,4-Dichlorobenzene.* ECHA/RAC/RES-O-0000003486-69-01/F. ECHA/SEAC/RES-O-0000003486-69-02/F. Available at: http://echa.europa.eu/documents/10162/7b884aff-2852-40c3-8e7a-642ca92ca425 (accessed December 8, 2013).

[66] Harris, J., Coons, S., Byrne, M., Fiksel, J., Goyer, M., Wagner, J., Wood, M., and Moss, K. 1981. *An Exposure and Risk Assessment for Dichlorobenzenes.* EPA-440/4-81-019.

[67] US EPA. 1982. *Occurrence of Dichlorobenzene in Drinking Water, Food and Air.* PB95-183075.

[68] US EPA. 1978. *Initial Report of the TSCA Interagency Testing Committee to the Administrator, Environmental Protection Agency,* EPA 560-10-78/001.

[69] US EPA. 1986. *Certain Chlorinated Benzenes. Final Test Rule*. 51 FR 24657-24667. July 8.

[70] US EPA. 2004. *In Vitro Dermal Absorption Rate Testing of Certain Chemicals of Interest to the Occupational Safety and Health Administration. ACTION: Final Rule.* 69 FR 22402.

[71] US EPA. 2008. *Results of Testing - G020 Chlorinated Benzenes*. Available at: http://www.epa.gov/opptintr/chemtest/pubs/cbenze.pdf (accessed September 7, 2014).

[72] US EPA. 2014. *TSCA Work Plan for Chemical Assessments: 2014 Update*. October. Available at: http://www.epa.gov/oppt/existingchemicals/pubs/TSCA_Work_Plan_Chemicals_2014_Update-final.pdf (accessed November 2, 2014).

[73] US EPA. 2010. *VCCEP Status p-Dichlorobenzene* (CAS No. 106-46-7). Web page last updated April 29, 2010. Available at: (accessed September 7, 2014).

[74] US EPA. 2011. *Evaluation Report: EPA's Voluntary Chemical Evaluation Program Did Not Achieve Children's Health Protection Goals.* Office of the Inspector General. Report No. 11-P-0379. Available at: http://www.epa.gov/oig/reports/2011/20110721-11-P-0379.pdf (accessed September 7, 2014).

[75] US EPA. 2008. *Reregistration Eligibility Decision for Para-dichlorobenzene*, Revised December 2008. EPA 738-R-07-010. Available at: http://www.epa.gov/opp00001/chem_search/reg_actions/reregistration/red_PC-061501_29-Dec-08.pdf (accessed September 7, 2014).

[76] CIRS. 2014. *Chinese Chemical Inventory Search System (CCISS)*. Available at: http://cciss.cirs-group.com/(accessed September 11, 2014).

[77] US EPA. 2014. *TRI EXPLORER - Releases: Trends Report*. Available at: http://ias-pub.epa.gov/triexplorer/release_trends?p_view=USYR&trilib=TRIQ1&sort=_VIEW_&sort_fmt=1&state=All+states&county=All+counties&chemical=000106467&industry=ALL&year=All+years&core_year=&tab_rpt=1&fld=AIRLBY&fld=E1&fld=E2&fld=E3&fld=E4&fld=E41&fld=E42&fld=E5&fld=E52&fld=E53&fld=E53A&fld=E53B&fld=E54&fld=E51&fld=E51A&fld=E51B&fld=TSFDSP&fld=TSFDSP&fld=m10&fld=m41&fld=m62&fld=potwmetl&fld=m71&fld=m81&fld=m82&fld=m72&fld=m63&fld=m64&fld=m65&fld=m66&fld=m67&fld=m73&fld=m79&fld=m90&fld=m94&fld=m99&fld=RELLBY (accessed September 5, 2014).

[78] McCarthy, M.C., Hafner, H.R., Chinkin, L.R., and Charrier, J.G. 2007. Temporal variability of selected air toxics in the United States. *Atmos. Environ.* 41(34): 7180–7194.

[79] Hoff, J.T., Wania, F., Mackay, D., and Gillham, R. 1995. Sorption of nonpolar organic vapors by ice and snow. *Environ. Sci. Technol.* 29(8): 1982–1989.

[80] Alaska Department of Environmental Conservation. 2008. *Fish Monitoring Program: Analysis of Organic Contaminants*. Available at: http://dec.alaska.gov/eh/docs/vet/Fish/Organic%20Data%202008.pdf (accessed September 30, 014).

[81] Morgenstern, L. 1987. *Knowledge Preconditions for Actions and Plans.* IJCAI: 867-874. Available at: http://ijcai.org/Past%20Proceedings/IJCAI-87-VOL2/PDF/064.pdf (accessed September 6, 2014).

[82] Trione, E.J. 1982. Dwarf bunt of wheat and its importance in international wheat trade. *Plant Diseases*. 66(11): 1083–1088.

[83] Wiese, M.V. and Smith, N.A. 1978. *Common Bunt Again a Threat to Wheat. Extension,* Bulletin E-1178. Cooperative Extension Service, Michigan State University.

[83] Yersin, H. 1947. Hexachlorobenzene, a synthetic organic product utilised in the control of wheat bunt. *Rev. Appl. Mycol.*, 26: 150–150.

[84] Taylor, E.L. 1945. Acaricidal property of a new insecticide, hexachlorobenzene. *Nature*. 155: 393–394.

[85] Mumma, C.E. and Lawless, E.W. 1975. *Survey of Industrial Processing Data, Task-1, Hexachlorobutadiene and Hexachlorobenzene Pollution from Chlorocarbon Processing*. EPA 560/3-75-003.

[86] Carpenter, C., Schweer, G., Stinnett, G., and Gabel, N. 1986. *Exposure Assessment for Hexachlorobenzene*. Prepared for EPA, Report No. EPA 560/5-86-019.

[87] US EPA. 1973. *Environmental Contamination from Hexachlorobenzene*. EPA 540/6-76-014. Office of Toxic Substances.

[88] Jones, K., Barber, J., and Sweetman, A. 2005. *Hexachlorobenzene-Sources, Environmental Fate and Risk Characterisation*. European Chlorinated Solvent Association (Euro Chlor).

[89] Wang, G., Lu, Y., Han, J., Luo, W., Shi, Y., Wang, T., and Sun, Y. 2010. Hexachlorobenzene sources, levels and human exposure in the environment of China. *Environ. Int*. 36(1): 122–130.

[90] ATSDR (Agency for Toxic Substances and Disease Registry). 1990. *Toxicological Profile for Hexachlorobenzene*. United States Public Health Service. 97 pp.

[91] Gouin, T., Mackay, D., Jones, K.C., Harner, T., and Meijer, S.N. 2004. Evidence for the "grasshopper" effect and fractionation during long-range atmospheric transport of organic contaminants. *Environ. Pollution*. 128: 139–148.

[92] Barber, J.L., Sweetman, A.J., Van Wijk, D., and Jones, K.C. 2005. Hexachlorobenzene in the global environment: Emissions, levels, distribution, trends and processes. *Sci. Total Environ*. 349(1): 1–44.

[93] Melis, R. 1954. Research on the toxic action of hexachlorobenzene (C6Cl6) on warm-blooded animals. *Nuovi Annali D'igiene e Microbiologia*, 6(5): 361–367.

[94] US General Accounting Office. 1975. Report to the Congress by the Comptroller General of the United States. *Federal Pesticide Registration Program: Is It Protecting The Public And The Environment Adequately From Pesticide Hazards*? RED-76-42.

[95] Courtney, K.D. 1979. Hexachlorobenzene (HCB): A review. *Environ. Res*. 20(2): 225–266.

[96] Dewdney, J. 2014. Turkey Under the Democrats, 1950–60. *Encyclopedia Britannica*. Last updated September 2, 2014. Available at: www.brittanica.com (accessed October 4, 2014).

[97] Cam, C. and Nigogosyan, G. 1963. Acquired toxic porphyria cutanea tarda due to hexachlorobenzene: Report of 348 cases caused by this fungicide. *JAMA*. 183(2): 88–91.

[98] Peters, H., Cripps, D., Göcmen, A., Bryan, G., Ertürk, E., and Morris, C. 1987. Turkish epidemic hexachlorobenzene porphyria a 30-year study. *Ann. New York Acad. Sci*. 514(1): 183–190.

[99] Agency for Toxic Substances and Disease Registry (ATSDR). 2013. *Toxicological Profile for Hexachlorobenzene*. (Draft for Public Comment) Atlanta, GA: US Department of Health and Human Services, Public Health Service. Available at: http://www.atsdr.cdc.gov/toxprofiles/tp.asp?id=627&tid=115 (accessed September 2014).

[100] Ockner, R.K. and Schmid, R. 1961. Acquired porphyria in man and rat due to hexachlorobenzene intoxication. *Nature*. 189:499.

[101] Jarrell, J.F., Gocmen, A., Akyol, D., and Brant, R. 2002. Hexachlorobenzene exposure and the proportion of male births in Turkey 1935–1990. *Reprod.Toxicol* 16(1): 65–70.

[102] Anonymous. 1996. *Extension Toxicology Network Pesticide Information Profiles. Hexachlorobenzene.* A Pesticide Information Project of Cooperative Extension Offices of Cornell University, Oregon State University, the University of Idaho, and the University of California at Davis and the Institute for Environmental Toxicology, Michigan State University. Major support and funding was provided by the USDA/Extension Service/National Agricultural Pesticide Impact Assessment Program. Available at: http://extoxnet.orst.edu/pips/hexachlo.htm (accessed October 5, 2014).

[103] IARC. 2001. *Monographs on the Evaluation of Carcinogenic Risks to Humans.* Volume 79. Some Thyrotropic Agents. pp. 493–568. Available http://monographs.iarc.fr/ENG/Monographs/vol79/index.php (accessed October 7, 2014).

[104] Euro Chlor. 2002. *Risk Assessment for the Marine Environment, OSPARCOM Region - North Sea. Hexachlorobenzene.* Available at: http://www.eurochlor.org/media/49363/8-11-4-15_marine_ra_hcb.pdf (accessed October 11, 2014).

[105] US GAO, H. 2009. Bioaccumulation of hexachlorobenzene in *Eisenia foetida* at different aging stages. *J. Environ. Sci.* 21(7): 948–953.

[106] United Nations Economic Commission for Europe. Undated. *Convention on Long-range Transboundary Air Pollution.* Available at: http://www.unece.org/fileadmin/DAM/env/lrtap/full%20text/1979.CLRTAP.e.pdf (accessed October 11, 2014).

[107] *PROTOCOL TO THE 1979 CONVENTION ON LONG-RANGE TRANSBOUNDARY AIR POLLUTION ON PERSISTENT ORGANIC POLLUTANTS.* Available at: http://www.unece.org/fileadmin/DAM/env/lrtap/full%20text/1998.POPs.e.pdf (accessed October 11, 2014).

[108] United Nations Treaty Collection. 2014. *Status of Treaties - CHAPTER XXVII ENVIRONMENT.* 1.g Protocol to the 1979 Convention on Long-Range Transboundary Air Pollution on Persistent Organic Pollutants. Available at: https://treaties.un.org/Pages/ViewDetails.aspx?src=TREATY&mtdsg_no=XXVII-1 g&chapter=27&lang=en (accessed October 11, 2014).

[109] Secretariat of the Stockholm Convention. 2008. *History of the Negotiations of the Stockholm Convention.* Available at: http://chm.pops.int/TheConvention/Overview/History/Overview/tabid/3549/Default.aspx (accessed October 11, 2014).

[110] Secretariat of the Stockholm Convention. Undated. *Status of Ratifications.* Available at: http://chm.pops.int/Countries/StatusofRatifications/tabid/252/Default.aspx#a-note-1 (accessed October 11, 2014).

[111] Secretariat of the Stockholm Convention. 2010. *Stockholm Convention on Persistent Organic Pollutants (POPs) as amended in 2009.* Text and Annexes.

[112] Blackwood, T.R. and Sipes, T.G. 1979. *Status Assessment of Toxic Chemicals: Hexachlorobenzene.* Prepared for US EPA Office of Research and Development. EPA-600/2-79-210g.

[113] The Council of European Communities. 1978. *Council Directive 79/117/EEC of 21 December 1978 Prohibiting the Placing on the Market and Use of Plant Protection Products Containing Certain Active Substances.* Available at: http://eur-lex.europa.eu/legal-content/EN/TXT/?uri=CELEX:31979L0117 (accessed October 11, 2014).

[114] The Council of European Communities. 1997. *Directive 97/56/EC of the European Parliament and of the Council of 20 October 1997.* Available at: http://eur-lex.europa.eu/legal-content/EN/TXT/?uri=CELEX:31997L0056 (accessed October 11, 2014).

[115] *Regulation (EC) No 850/2004 of the European Parliament and of the Council of 29 April 2004 on Persistent Organic Pollutants and Amending Directive* 79/117/EEC. Available at: http://eur-lex.europa.eu/legal-content/en/ALL/;ELX_SESSIONID=SDKTJTpLWRRhz60P9B942V3ygCJQ5Ln19CKnnZx6hVJvJTG9Wtq4!-398220743?uri=CELEX:32004R0850 (accessed October 11, 2014).
[116] *Regulation (EC) No 1907/2006 of the European Parliament and of the Council of 18 December 2006 Concerning the Registration, Evaluation, Authorisation and Restriction of Chemicals (REACH), Establishing a European Chemicals Agency, Amending Directive 1999/45/EC and Repealing Council Regulation (EEC) No 793/93 and Commission Regulation (EC) No 1488/94 as well as Council Directive 76/769/EEC and Commission Directives* 91/155/EEC, 93/67/EEC, 93/105/EC and 2000/21/EC. Available at: http://eur-lex.europa.eu/LexUriServ/LexUriServ.do?uri=CELEX:32006R1907:EN:NOT (accessed November 11, 2013).
[117] SEPA. 2007. *The People's Republic of China National Implementation Plan for the Stockholm Convention on Persistent Organic Pollutants. Part I - Implementation Plan of the Chinese Central Government for the Stockholm Convention on Persistent Organic Pollutants.* Available at: http://www.pops.int/documents/implementation/nips/submissions/China_NIP_En.pdf (accessed October 11, 2014).
[118] United Nations. 2014. *Commodity Trade Statistics Database*. Available at: http://comtrade.un.org/data/(accessed October 2014).
[119] UNEP. 2009. *Global Monitoring Report Under the Global Monitoring Plan for Effectiveness Evaluation. Stockholm Convention on Persistent Organic Pollutants.* UNEP/POPS/COP.4/33. Available at: http://chm.pops.int/Implementation/GlobalMonitoringPlan/MonitoringReports/tabid/525/Default.aspx (accessed October 12, 2014).
[120] Song, S., Ma, J., Tian, Q., Tong, L., and Guo, X. 2013. Hexachlorobenzene in human milk collected from Beijing, China. *Chemosphere*. 91(2): 145–149.
[121] Wilson, S., Hung, H., Katsoyiannis, A., Kong, D., van Oostdam, J., Rigét, F., and Bignert, A. 2014. *Trends in Stockholm Convention persistent organic pollutants (POPs) in arctic air, human media and biota*. AMAP Technical Report No. 7 (2014). Arctic Monitoring and Assessment Programme (AMAP), Oslo, 54 pp.
[122] Blood, N.O.F. 1948. Pesticides: Chemical contaminants of foods. *JAMA*. 137(18): 1604–1605.
[123] Schier, J.G., Sejvar, J.J., Lutterloh, E., Likaka, A., Katsoudas, E., Karaseva, Y.D., and Monroe, S. (2012). Consumption of pesticide-treated wheat seed by a rural population in Malawi. *J. Exposure Sci. Environ.Epidemiol*. 22(6): 569–573.
[124] Browne, M.A., Crump, P., Niven, S.J., Teuten, E., Tonkin, A., Galloway, T., and Thompson, R. 2011. Accumulation of microplastic on shorelines worldwide: Sources and sinks. *Environ. Sci. Technol*. 45(21): 9175–9179.
[125] Van Cauwenberghe, L., Vanreusel, A., Mees, J., and Janssen, C.R. 2013. Microplastic pollution in deep-sea sediments. *Environ. Pollution*. 182: 495–499.
[126] Cole, M., Lindeque, P., Halsband, C., and Galloway, T.S. 2011. Microplastics as contaminants in the marine environment: A review. *Marine Pollution Bull*. 62(12): 2588–2597.
[127] Andrady, A.L. 2011. Microplastics in the marine environment. *Marine Pollution Bull*. 62(8): 1596–1605.
[128] Beach, W.J. 1972. *Skin Cleaner*. Sugar Beet Products Company, Saginaw, Michigan, assignee. Patent 3645904. 29 Feb. 1972.

[129] Bara, I. and Melllul, M. 1997. *Deep-Cleansing Composition Containing Particles of Expanded Polymer*. L'Oreal, Paris, France, Assignee. US Patent 5,679,326. 21 October 1997.

[130] Zitko, V. and Hanlon, M. 1991. Another source of pollution by plastics: Skin cleaners with plastic scrubbers. *Marine Pollution Bull.* 22(1): 41–42.

[131] Gregory, M.R. 1996. Plastic 'scrubbers' in hand cleansers: A further (and minor) source for marine pollution identified. *Marine Pollution Bull.* 32(12): 867–871.

[132] Fendall, L.S. and Sewell, M.A. 2009. Contributing to marine pollution by washing your face: Microplastics in facial cleansers. *Marine Pollution Bull.* 58(8): 1225–1228.

[133] Chua, E., Shimeta, J., Nugegoda, D., Morrison, P.D., and Clarke, B.O. 2014. *Assimilation of Polybrominated Diphenyl Ethers from Microplastics by the Marine Amphipod*, Allorchestes Compressa. Environmental Science & Technology.

[134] Connor, S. and Swinburne, Z. 2014. Beauty brands pledge to end use of microbeads in their products. *The Independent*. May 25. Available at: http://www.independent.co.uk/news/science/beauty-brands-pledge-to-end-use-of-microbeads-in-their-products-9431984.html (accessed June 4, 2014).

[135] International Campaign against Microbeads in Cosmetics. Undated. *Product Lists*. Available at: http://www.beatthemicrobead.org/en/product-lists (accessed October 31, 2014).

[136] Nalbone, J. 2014. *Unseen Threat: How Microbeads Harm New York Waters, Wildlife, Health and Environment*. From the Office of New York State Attorney General Eric T. Schneiderman. Available at: http://ag.ny.gov/pdfs/Microbeads_Report_5_14_14.pdf (accessed August 1, 2014).

[137] Tchobanoglous, G. and Burton, F.L. 1991. *Wastewater Engineering: Treatment, Disposal and Reuse*, 3rd edition. New York: Metcalf & Eddy; McGraw-Hill, Chapter 9.

[138] Hogue. 2013. Microplastic beads pollute Great Lakes. *Chem. Eng. News*. 91(37): 23–25. Available at: http://cen.acs.org/articles/91/i37/Microplastic-Beads-Pollute-Great-Lakes.html (accessed August 1, 2014).

[139] Sahagun, L. 2014. Microbeads a major problem in L.A. River. *Los Angeles Times*. January 25. Available at: http://www.latimes.com/science/la-me-0126-microbeads-20140126-story.html (accessed August 1, 2014).

[140] Hammer, J., Kraak, M., and Parsons, J.R. 2012. Plastics in the marine environment: The dark side of a modern gift. *Reviews of Environmental Contamination and Toxicology*. New York: Springer, pp. 1–44.

[141] Eriksen, M., Maximenko, N., Thiel, M., Cummins, A., Lattin, G., Wilson, S., and Rifman, S. 2013. Plastic pollution in the South Pacific subtropical gyre. *Marine Pollution Bull.* 68(1): 71–76.

[142] Cózar, A., Echevarría, F., González-Gordillo, J.I., Irigoien, X., Úbeda, B., Hernández-León, S., and Duarte, C.M. 2014. Plastic debris in the open ocean. *Proc. Nat. Acad. Sci.* 111(28): 10239–10244 and supplemental material.

[143] Alldredge, A.L. and Silver, M.W. 1988. Characteristics, dynamics and significance of marine snow. *Prog.Oceanog.* 20(1): 41–82.

[144] Wright, S.L., Thompson, R.C., and Galloway, T.S. 2013. The physical impacts of microplastics on marine organisms: A review. *Environ. Pollution*. 178: 483–492.

[145] Engler, R. 2012. The complex interaction between marine debris and toxic chemicals in the ocean. *Environ. Sci. Technol.* 46: 12302–12315.

[146] Teuten, E.L., Rowland, S.J., Galloway, T.S., and Thompson, R.C. 2007. Potential for plastics to transport hydrophobic contaminants. *Environ. Sci. Technol.* 41(22): 7759–7764.

[147] Rios, L.M., Jones, P.R., Moore, C., and Narayan, U.V. 2010. Quantitation of persistent organic pollutants adsorbed on plastic debris from the Northern Pacific Gyre's "eastern garbage patch." *.J. Environ. Monitor.* 12(12): 2226–2236.

[148] Gouin, T., Roche, N., Lohmann, R., and Hodges, G. 2011. A thermodynamic approach for assessing the environmental exposure of chemicals absorbed to microplastic. *Environ. Sci. Technol.* 45(4): 1466–1472.

[149] Teuten, E.L., Saquing, J.M., Knappe, D.R., Barlaz, M.A., Jonsson, S., Björn, A., and Takada, H. 2009. Transport and release of chemicals from plastics to the environment and to wildlife. *Philosoph. Trans. Roy. Soc. B: Biol. Sci.*, 364(1526): 2027–2045.

[150] Zarfl, C. and Matthies, M. 2010. Are marine plastic particles transport vectors for organic pollutants to the Arctic? *Marine Pollution Bull.* 60(10):1810–1814.

[151] Moore, C.J., Moore, S.L., Leecaster, M.K., and Weisberg, S.B. (2001). A comparison of plastic and plankton in the North Pacific central gyre. *Marine Pollution Bull.* 42(12): 1297–1300.

[152] Engler, R.E. 2012. The complex interaction between marine debris and toxic chemicals in the ocean. *Environ. Sci. Technol.* 46(22): 12302–12315.

[153] Bakir, A., Rowland, S.J., and Thompson, R.C. 2014. Enhanced desorption of persistent organic pollutants from microplastics under simulated physiological conditions. *Environ. Pollution.* 185: 16–23.

[154] Koelmans, A.A., Besseling, E., Wegner, A., and Foekema, E.M. 2013. Plastic as a carrier of POPs to aquatic organisms: A model analysis. *Environ. Sci. Technol.* 47(14): 7812–7820.

[155] Koelmans, A.A., Besseling, E., and Foekema, E.M. 2014. Leaching of plastic additives to marine organisms. *Environ. Pollution.* 187:49–54.

[156] Browne, M.A., Niven, S.J., Galloway, T.S., Rowland, S.J., and Thompson, R.C. 2013. Microplastic moves pollutants and additives to worms, reducing functions linked to health and biodiversity. *Curr. Biol.* 23(23): 2388–2392.

[157] Besseling, E., Wegner, A., Foekema, E.M., van den Heuvel-Greve, M.J., and Koelmans, A.A. 2012. Effects of microplastic on fitness and PCB bioaccumulation by the lugworm Arenicola marina (L.). *Environ. Sci. Technol.* 47(1): 593–600.

[158] Ryan, P.G., Connell, A.D., and Gardner, B.D. 1988. Plastic ingestion and PCBs in seabirds: Is there a relationship? *Marine Pollution Bull.* 19(4): 174–176.

[159] Yamashita, R., Takada, H., Fukuwaka, M.A., and Watanuki, Y. 2011. Physical and chemical effects of ingested plastic debris on short-tailed shearwaters, Puffinus tenuirostris, in the North Pacific Ocean. *Marine Pollution Bull.* 62(12): 2845–2849.

[160] Tanaka, K., Takada, H., Yamashita, R., Mizukawa, K., Fukuwaka, M.A., and Watanuki, Y. 2013. Accumulation of plastic-derived chemicals in tissues of seabirds ingesting marine plastics. *Marine Pollution Bull.* 69(1): 219–222.

[161] Library of Congress. 2014. *H.R.4895— Microbead-Free Waters Act of 2014* (Introduced in House - IH). Available at: http://thomas.loc.gov/cgi-bin/thomas (accessed August 10, 2014).

[162] Zook,M. 2014. Illinois bans microbeads in personal care products: A growing trend in public and private policy. *Chemical Watch*, July 31. Available at: http://chemicalwatch.com/20743/illinois-bans-microbeads-in-personal-care-products (accessed July 31, 2014).

[163] ECHA. 2012. *Guidance for Monomers and Polymers. Version 2.0.* Pp. 14. Available at: http://echa.europa.eu/documents/10162/13632/polymers_en.pdf (accessed August 10, 2014).

[164] ECHA. 2011. *Guidance in a Nutshell: Requirements for Substances in Articles.* Available at: http://echa.europa.eu/documents/10162/13632/nutshell_guidance_articles2_en.pdf (accessed August 10, 2014).

[165] Plastic Soup Foundation and Stichting De Noordzee. Undated. *INTERNATIONAL CAMPAIGN AGAINST MICROBEADS IN COSMETICS.* Available at: http://www.beatthemicrobead.org/en/politics (accessed August 10, 2014).

[166] Yang, Q. and Wen, E. 2011. *Polymer Regulations and Polymer Notifications Update Report. CIRS – Chemical Inspection and Regulation Service.* Available at: http://www.cirs-reach.com/Polymer_Regulations_New_Polymer_Notificatios_Update_Report.pdf (accessed August 16, 2014).

[167] Plastic Soup and DeNoordzee. *Beat the Microbead App*. Available at: http://get.beatthemicrobead.org/(accessed August 10, 2014).

[168] Connor, S. and Swinburne, Z. 2014. Beauty brands pledge to end use of microbeads in their products. *The Independent*. May 25. Available at: http://www.independent.co.uk/news/science/beauty-brands-pledge-to-end-use-of-microbeads-in-their-products-9431984.html (accessed August 10, 2014).

[169] Nearing, B. 2014. State moves toward microbead ban, industry urges delay. *Times Union*. May 14. Available at: http://www.timesunion.com/business/article/State-moves-toward-microbead-ban-industry-urges-5478990.php (accessed August 10, 2014).

[170] US EPA. 2012. *The Great Lakes: An Environmental Atlas and Resource Book.* Chapter 1 - Introduction: The Great Lakes and Chapter Three - People and the Great Lakes, Web page last updated June 25, 2012. Available at: http://www.epa.gov/greatlakes/atlas/glat-ch3.html (accessed August 1, 2014).

[171] Chaskey, E., Hirsch, T., Drake, T., Ehmann, K., and Chu, Y. 2014. *Micro-plastic Pollution: A Comparative Survey of Wastewater Effluent in New York. Center for earth and Environmental Science Student Posters. Book 8.* Available at: http://digitalcommons.plattsburgh.edu/cees_student_posters/8 (accessed August 1, 2014).

[172] Eriksen, M., Mason, S., Wilson, S., Box, C., Zellers, A., Edwards, W., and Amato, S. 2013. Microplastic pollution in the surface waters of the Laurentian Great Lakes. *Marine Pollution Bull.* 77(1): 177–182.

[173] Ohio Department of Natural Resources, 2014. *Ohio's Lake Erie Fisheries 2013.* Available at: http://wildlife.ohiodnr.gov/portals/wildlife/pdfs/fishing/Lake%20ErieStatus2013.pdf (accessed August 1, 2014).

[174] Johnston, C. 2013. Personal grooming products may be harming Great Lakes marine life. *Sci. Amer.* June 25, 2013. Available at: http://www.scientificamerican.com/article/microplastic-pollution-in-the-great-lakes/(accessed August 1, 2014).

[175] Thompson, R.C., Olsen, Y., Mitchell, R.P., Davis, A., Rowland, S.J., John, A.W., and Russell, A.E. 2004. Lost at sea: Where is all the plastic? *Science.* 304(5672): 838–838.

[176] Carpenter, E.J., Anderson, S.J., Harvey, G.R., Miklas, H.P., and Peck, B.B. 1972. Polystyrene spherules in coastal waters. *Science.* 178(4062): 749–750.

[177] Colton, J.B., Knapp, F.D., and Burns, B.R. 1974. Plastic particles in surface waters of the northwestern Atlantic. *Science.* 185(4150): 491–497.

[178] Hays, H. and Cormons, G. 1974. Plastic particles found in tern pellets, on coastal beaches and at factory sites. *Marine Pollution Bull.* 5(3), 44–46.
[179] Habib, D., Locke, D.C., and Cannone, L.J. 1998. Synthetic fibers as indicators of municipal sewage sludge, sludge products, and sewage treatment plant effluents. *Water, Air, Soil Pollution*. 103(1–4): 1–8.

5

Conclusions

This book began with two deliberately naïve questions: "How is it that man-made chemicals are found in our bodies and in even remote parts of our planet?" and "Does it matter?" The answers to those questions reflect our ability to predict, understand, and regulate the consequences of using a chemical substance through its life cycle.

The first part of the answers has to do with time. The modern chemical industry was far advanced before we began to ask those questions seriously or have the tools to answer them. Beginning in the early 1700s with the commercial production of acids and alkalis, the industry has grown steadily, with the number and complexity of chemicals entering the market rising sharply in the late 1800s and in the post-World War II era. By the time we began to enact chemical control laws in the 1970s, chemists had created over 3,000,000 substances. Approximately 50,000 of those chemicals were on the market, albeit perhaps only a few thousand represented the majority of sales. We got ahead of ourselves. Although we could see the benefits to using new chemicals with respect to their functions and in the jobs that manufacturing created, we lacked all but the crudest of tools to understand and regulate the consequences of their manufacture and use until relatively recently in the 300-year history of the modern chemical industry.

Those tools, whether legislative or technical, have been evolving over the past 40 years. One consequence of the evolution has been that decisions to regulate or restrict a chemical have sometimes taken decades. The European Commission began, for example, to consider the risks from exposure to 1,4-dichlorobenzene in the 1990s but restrictions were not promulgated until 2014. In the time that it can take for authorities to make careful balanced decisions, chemicals may be released and transported through the environment. And sometimes, as in the case of 1,4-dichlorobenzene, during the period of evaluation market factors can change the uses of a substance and the potential for exposure and risk.

If the first part of the answers to the questions that frame this book relates to *time*, the next consideration is the balance implicit in risk management. Risk management, as we discussed in Chapter 2, is a "decision-making process involving considerations of political, social, economic, and technical factors with relevant risk assessment information relating to a hazard so as to develop, analyse, and compare regulatory and non-regulatory options and to select and implement appropriate regulatory response to that hazard" [1]. Or put more simply, decision makers—whether in the halls of Congress or the aisles of a store—must balance the benefits of chemical use against the costs and possible consequences. Those choices reflect the values of the day, which may not be shared by previous or subsequent generations. The US Toxic Substances Control Act (TSCA), legislated during the oil crisis and devastating loss of jobs in the early 1970s, reflected the understandable concern that strict chemical control regulation would further weaken the economy. Some 40 years later legislators debated many of the provisions of TSCA but agreed that the requirement to regulate chemicals "in such a manner as not to impede unduly or create unnecessary economic barriers to technological innovation" should stand [2]. For another perspective, consider the shift in consumers' risk-management decisions implicit in the following trend. According to a report by the US Department of Agriculture, sales of organic food in the United States grew from approximately $11M US in 2004 to an estimated $35M US in 2013 [3]. During that decade many consumers considered "economic and technical factors with relevant risk assessment information relating to a hazard" and chose to purchase more expensive organic food over conventionally farmed food, a risk-management decision that differed from that of a previous generation.

Our risk-management decisions reflect our field of vision, beginning with the hazards we consider. Our knowledge of hazards depends on the tests we run. Logically enough, early toxicity testing focused on acute effects. Although the suite of available tests has grown to include a variety of acute and chronic effects, we still wrestle with how and when to characterize such subtle effects as neurotoxicity and endocrine disruption and lack the tools to evaluate the effects of exposure to multiple chemicals. Our tools for exposure and risk assessment require us to make many assumptions to approximate real-world processes. Those assumptions reflect both scientific data and subjective judgments. Consider, for example, that dose–response factors for a single chemical can vary by orders of magnitude under different regulatory programs, which can lead to different risk-assessment conclusions and risk-management decisions. In consequence of the number and variability of assumptions, our view of the possible consequences of chemical usage may be blurred.

Our field of vision also reflects the aperture through which we view risks. Efforts to characterize risk may necessarily, due to a scientific or regulatory focus on a single endpoint, miss a larger consequence. Consider the story of PFOS related in Chapter 2. Discharges from wastewater treatment plants around the world have contained PFOS at hundreds of nanograms per liter, concentrations orders of magnitude below the levels at which aquatic toxicity would be expected. One might conclude, based on that assessment of ecological risks, that the discharges did not merit concern or the chemical deserve regulation. But the PFOS in those discharges has accumulated in ocean waters and been transported around the world. Looking through that wider aperture, the United Nations Environment Programme has characterized PFOS as a Persistent Organic Pollutant that should be banned based on its bioaccumulation in animals in the Arctic, far from any sources. That risk-management judgment differs from one that might be made based on a narrower focus.

Risk-management protocols sometimes single out persistent, bioaccumulative, and toxic (PBT) compounds. The Stockholm Convention and related national legislation have effected bans on PBT compounds such as hexachlorobenzene. But even here, the risk-management vision is sometimes unclear. The criteria used to define PBT compounds are somewhat arbitrary [4]. Further, they differ between jurisdictions, particularly with respect to the threshold values for characterizing a compound as persistent, bioaccumulative, or toxic. Most regulatory programs lack one important criterion, the half-life in air, that would identify PBT compounds that might migrate beyond national borders. The US EPA does not consider the half-life in air when assessing a new chemical substance under TSCA, but oddly enough uses that criterion in Toxic Release Inventory reporting [5,6]. Under the Regulation on Registration, Evaluation, Authorisation and Restriction of Chemicals, the European Union does not factor the half-life in air into PBT assessments [7]. Exceptions are rare: Canada, for example, does consider a relatively long half-life in air as a sign of persistence [7]. In part due to the variations in the criteria for defining PBT compounds, the number of substances recognized as PBT varies widely between jurisdictions but typically numbers less than 100. A recent screening of chemical inventories indicates that between 100 and 1,000 PBT compounds may actually be in commerce [4]. And so it seems that we lack a clear view of which substances may be most hazardous. We also lack a full understanding of the consequences of the release of PBT compounds into the environment. As illustrated by the results of the simple fugacity models shown in Chapter 4, persistent compounds tend to partition to soil and sediment. Yet our ecotoxicity test regimes focus on aquatic plants and animals. Only rarely are chemicals brought to market with test data on the effects of exposure by terrestrial organisms such as earthworms, or to benthic organisms. Our technical understanding of the nature of PBT compounds as expressed through national legislation and our characterization of the consequences of ecological exposures factor heavily into risk-management decisions regarding PBT compounds, and these flaws in our understanding may lead to short-sighted decisions.

The third consideration in our answers to the questions, "How is it that man-made chemicals are found in our bodies and in even remote parts of our planet?" and "Does it matter?" is our ability to understand the consequences of our risk-management decisions after they are made and to use that information appropriately. We often lack the data to assess the results of taking action. Sometimes the data suggest that the results may not be as intended. The intended consequence of banning the production and trade of hexachlorobenzene starting nearly 30 years ago, for example, was to reduce the risk from environmental contamination. Monitoring data indicate that environmental concentrations in many areas are decreasing, although the concentrations in environmental samples from the Arctic and blood samples from people living in that region are not. Trade data suggest that global commerce in hexachlorobenzene and DDT, both banned, has not significantly abated.

This attempt to answer two simple questions about the consequences of the manufacture and use of chemicals has traversed a broad array of topics, from the precepts of a chemical engineer's mass balance to a toxicologist's assessment of the consequences of exposures, and on through to the regulatory regimes that interpret and apply the results. The answers, which at heart relate to how we perceive and manage threats, lie in the domain of philosophers as much as in the realm of science or engineering. They reflect our human ability to foresee the consequences of our actions, to judge those consequences as acceptable or not, and to take effective action. Perhaps the best way to close this exploration is with a quotation attributed to Albert Einstein:

> Learn from yesterday, live for today, hope for tomorrow. The important thing is not to stop questioning.

References

[1] OECD. 2003. *Descriptions of Selected Key Generic Terms used in Chemical Hazard/Risk Assessment*. OECD Series on Testing and Assessment, Number 44. ENV/JM/MONO(2003)15. Available at: http://search.oecd.org/officialdocuments/publicdisplaydocumentpdf/?cote=ENV/JM/MONO(2003)15&docLanguage=En (accessed September 14, 2013).

[2] Bergeson & Campbell, PC. 2014. *Regulatory Developments: Summary of Changes in the Chemical Safety Improvement Act (CSIA, S. 1009) and the Boxer TSCA Version Released September 18, 2014*. Available at: http://www.lawbc.com/regulatory-developments/entry/summary-of-changes-in-the-chemical-safety-improvement-actcsia-s.-1009-and-t (accessed November 2, 2014).

[3] Greene, C. 2013. *Growth Patterns in the U.S. Organic Industry*. U.S. Department of Agriculture Economic Research Service. Available at: http://www.ers.usda.gov/amber-waves/2013-october/growth-patterns-in-the-us-organic-industry.aspx (accessed October 22, 2014).

[4] Abelkop, A.D.K., Bergkamp, L., Brooks, B.W., Gergely, A., Graham, J.D., Gray, G., van Leeuwen, K., Marchant, G.E., Mueller, M.L., Royer, T.V., and Vighi, M. 2013. *Scientific and Policy Analysis of Persistent, Bioaccumulative, and Toxic Chemicals: A Comparison of Practices in Asia, Europe, and North America*. The Report of a Consensus Panel. Available at: http://www.indiana.edu/~spea/faculty/pdf/scientific_policy_analysis_of_persistent_bioaccumulative_and_toxic_chemicals_PBT_.pdf (accessed October 23, 2014).

[5] Category for persistent, bioaccumulative, and toxic new chemical substances *Federal Register*: November 4, 1999 (Volume 64, Number 213), pp. 60194–60204.

[6] Environmental Health Analysis Center. 2012. *Criteria used by the PBT Profiler.* Developed by the Environmental Health Analysis Center under contract to the Office of Chemical Safety and Pollution Prevention, U.S. Environmental Protection Agency. Ver 2.000. Last updated September 4, 2012. Available at: http://www.pbtprofiler.net/criteria.asp (accessed October 23, 2014).
[7] ECHA. 2012. *Guidance on Information Requirements and Chemical Safety Assessment Chapter R.11: PBT Assessment.* Table R. 11-1: PBT and vPvB criteria according to Annex XIII. Available at: http://echa.europa.eu/documents/10162/13632/information_requirements_r11_en.pdf (accessed October 22, 2014).
[8] *Persistence and Bioaccumulation Regulations (SOR/2000-107).* Available at: http://laws-lois.justice.gc.ca/eng/regulations/SOR-2000-107/index.html (accessed October 23, 2014).

Index

Note: Page numbers ending in "f" refer to figures. Page numbers ending in "t" refer to tables.

www.ingramcontent.com/pod-product-compliance
Lightning Source LLC
Chambersburg PA
CBHW060745150926
51501CB00003B/99
* 9 7 8 1 4 8 2 2 2 3 2 9 3 *